THE COW
WITH EAR TAG
#1389

KATHRYN GILLESPIE

The University of Chicago Press • Chicago and London

The University of Chicago Press, Chicago 60637
The University of Chicago Press, Ltd., London
© 2018 by The University of Chicago
Published 2018 Printed in the United States of America

27 26 25 24 23 22 21 20 19 18   1 2 3 4 5

ISBN-13: 978-0-226-58271-9(cloth)
ISBN-13: 978-0-226-58285-6(paper)
ISBN-13: 978-0-226-58299-3(e-book)
DOI: https://doi.org/10.7208/chicago/9780226582993.001.0001

Library of Congress Cataloging-in-Publication Data

Names: Gillespie, Kathryn (Kathryn A.), author.
Title: The cow with ear tag #1389 / Kathryn Gillespie.
Description: Chicago : The University of Chicago Press, 2018. | Includes
    bibliographical references and index.
Identifiers: LCCN 2018014000 | ISBN 9780226582719 (cloth : alk. paper) |
    ISBN 9780226582856 (pbk. : alk. paper) | ISBN 9780226582993 (e-book)
Subjects: LCSH: Animal welfare—Washington (State) | Animal welfare—
    California. | Dairy farming—Washington (State) | Dairy farming—
    California. | Animal welfare—Moral and ethical aspects. | Food
    animals—Psychological aspects.
Classification: LCC HV4765.W2 G55 2018 | DDC 179/.3—dc23
LC record available at https://lccn.loc.gov/2018014000

♾ This paper meets the requirements of ANSI/NISO Z39.48–1992
(Permanence of Paper).

THE COW

WITH EAR TAG

#1389

CONTENTS

FIGURE 1.1   Sadie, Animal Place (Grass Valley, CA)

1

SADIE

With rattlesnake guards wrapped around our feet and legs, Marji Beach and I tromped through the tall grass out into the field where the geriatric herd lives. Marji is the education director for Animal Place, a sanctuary for formerly farmed animals in Grass Valley, California. The sanctuary, situated on hundreds of acres, is part forest, part lush green field. Barns and other animal housing dot the landscape and a farmhouse serves as the visitor and education center at the front of the property. A large organic garden near the house is tended by sanctuary workers, producing enough to run a community-supported agriculture produce program for the surrounding community.

To the casual observer, Animal Place looks like the picturesque family farm one might imagine when thinking of animal agriculture: cows and pigs grazing the fields, chickens pecking in the yard, a productive vegetable garden growing enough to sustain the farm, with enough left over to sell; and rolling green fields and large old trees meeting a clear blue sky populated by white clouds. And yet, what's going on at Animal Place is different. The animals here are not farmed. As a sanctuary, it is a place dedicated to the care and rehabilitation of animals who have labored or experienced abuse or neglect on farms around the country. Animal Place is a place where animals who would otherwise be farmed for milk, meat, or eggs can live out their lives.

Marji Beach explained this to me—the mission of Animal Place—as we approached the geriatric herd. Several old cows and steers lounged in the field in the June morning sun. We approached slowly. Some of these animals had remained wary of human strangers despite the fact that many had spent years at the sanctuary learning to trust and love the sanctuary employees. Sadie, Elsa, and Howie lay in the tall grass not far from one another. Sadie was a Holstein (with the characteristic black-and-white patches), the most common breed of cow used for dairy in the United States. As we stood there, Marji told me what was known of Sadie's history.

Sadie was born and raised on a fairly large dairy farm housing approximately three thousand cows in the San Francisco Bay Area in central California. At the farm, her tail had been docked and her ears were tagged for identification. She was impregnated via artificial insemination at eighteen months old and then once a year every year after that. Her calves were taken away from her hours after birth, and she was milked three times a day until her productivity and reproductive capacity waned. At age five, Sadie developed a bad case of mastitis (a common result of any one of a number of pathogens that can easily infect the udders) and was sent to auction for slaughter with many other cows who had been used for dairy production. Sadie was bought at auction by a university veterinary teaching hospital, to be used as a teaching tool. She spent approximately twenty weeks (two academic quarters) at the school, where students used her to practice venipuncture (finding a vein in the neck and drawing blood) and rectal exams.

At the end of the twenty weeks, Sadie was going to be sent back to auction for slaughter, but a veterinary student intervened and contacted Animal Place to take her in. While at the university, Sadie did not receive medical treatment for her mastitis and the infection got worse, so the staff at Animal Place took her back to the teaching hospital for treatment. Sadie's mastitis case was severe and extremely painful. During treatment, which lasted nearly two years, her distrust of humans grew. Adding to this distrust, the staff at the teaching hospital had hurried her while loading her into a trailer and Sadie panicked and fell and broke her leg and hip—an injury from which it is difficult for even a young, healthy cow to recuperate.

When she returned to Animal Place, sanctuary employees also discovered that she was pregnant—something that went unnoticed at the veterinary teaching hospital. Not long after, Sadie gave birth to a

stillborn calf at the sanctuary. Because of the nature of dairy industry practices, she had never spent time with her newborn calves while
being used for milk production. At Animal Place, she spent several
hours with her dead calf, grooming him after he was born, before he
was buried at the sanctuary.

For a number of years after that, Sadie lived in the main herd at the
sanctuary and welcomed orphaned calves who were new to the sanctuary. As we stood there in the field, Marji told me that Sadie never
fully recovered from her experiences on the farm and at the teaching
hospital and remained wary of humans, but her caretakers believed
that she enjoyed living in community with the other cows and steers
and parenting orphaned calves new to the sanctuary. When she became too old and frail to live safely with the main herd (she could be
easily injured by younger, rambunctious steers), she was moved into
the geriatric herd with the other older cows and steers.

While we talked, Sadie relaxed and bowed her head for Marji to
scratch her neck and back. Flies were landing on her back, and she
stretched forward for Marji to brush them away and to scratch those
hard-to-reach places. Sadie's story could easily be read on her fourteen-
year-old body: in her docked tail that could not swat flies away from
biting her back, in the holes permanently punched in her ears from the
dairy farm's ear-tagging system, in the chronic limp from her leg/hip
injury, and in her wariness and distrust of strangers. As I stood there
watching Sadie and looking out across the field at Animal Place, I was
struck by the way in which Sadie's life was extraordinary: most cows
are not sold to veterinary teaching hospitals, and most cows do not
end up in sanctuaries. But it was precisely her extraordinary life history
that enabled a window into understanding the lasting effects of life on
a dairy farm, and in a teaching hospital.

ENCOUNTERING FARMED ANIMALS

A decade ago, when I first began thinking about the politics of food
production that would eventually, after a meandering journey, lead to
the research for this book, I imagined that cows raised for dairy lived

like Sadie did at Animal Place. This vision was rooted in my childhood. I had grown up in western Pennsylvania and had driven through the countryside and seen cows on dairy farms grazing in fields in front of red barns. As a kid, I would sit in the back seat of the car and rest my chin on the windowsill, watching as we whizzed by bucolic landscapes peppered with farmed animals. My father would announce matter-of-factly, "cows!" "goats!" "sheep!"—making sure we didn't miss seeing any of the animals as we passed. We lived in the heart of urban Pittsburgh, and these road trips out through the country were one of our only points of contact with farmed animals. My great-aunts lived in rural Virginia, and we would travel there in the summers to visit them; sometimes we would go down the road to a horse farm where my great-aunts knew the owners, and we would get to pet the horses. And now and then, we would stop on the side of the road on our travels and feed handfuls of grass or carrots to horses or cows over the fence and squeal as they nuzzled our hands with their noses looking for treats. To be that close to such a large, gentle animal was exhilarating. To look into the eyelashed eyes of a cow and feel her breath hot on my face as she snorted and shook her head was a feeling I remembered vividly for years.

Humans' relationships with other species enrich our lives in ways we may or may not readily acknowledge. Many humans engage in relationships of care and love with other animals—in the West, for instance, the dogs and cats with whom many people share their homes frequently become family members, often sharing their beds, their couches, and their emotional ups and downs. These animals are loved. They are worried over. Their thoughts and feelings are considered and wondered about. And they are grieved when they die. Although grieving the death or suffering of other species is sometimes dismissed ("It's *only* a dog—get over it already!"), those who have lived closely with, and loved, other animals understand the power and depth of these relationships and the hole they leave in our lives when they are gone.

These close relationships with animals are often limited to those with whom people share their homes and lives. But meeting farmed animal species and getting to know them on their own terms reveals that the nonhuman animal species that humans farm for food are not altogether different from the dogs and cats who are considered pets.[1] The first time I met a pig face to face, I was shocked by how much she seemed like a dog. Her name was Ziggy, and she was a large, three-legged, pink farm pig at Pigs Peace Sanctuary in Stanwood, Washington. I approached and she snorted at me and prodded my hand with her snout. I scratched her behind the ears and rubbed her back. With one impressive thump, she flopped over on her side on the ground and stuck her legs out. Judy Woods, the sanctuary director, said expectantly, "Well?"

"Well, what?" I asked, puzzled.

"She obviously wants you to scratch her belly!"

"Oh!" I immediately knelt and rubbed her belly with my hand.

"Use your finger nails," Judy instructed. And I did.

Ziggy closed her eyes and laid her head back on the grass. If I stopped scratching, even for a moment, she would raise her head and look at me, and Judy would say, "She didn't say you could stop."

Over years of volunteering at Pigs Peace, I've gotten to know many different pigs who live there — enormous Baily (blind at birth and living in the sanctuary's "special needs" area), old Betsy (rescued from a family farm where Animal Control found her, too weak to stand, resting her head on the body of one of her dead pen-mates to keep from drowning in the mud), and Honey (a piglet found with crushed hind legs on the floor of a pick-up truck when the driver was stopped for a DUI).

These animals each have stories and personalities of their own, with distinct likes and dislikes, histories, and emotional traumas they carry with them. To talk about the emotional worlds of nonhuman animals often draws the charge of anthropomorphism, which means attributing what we consider to be distinctly human characteristics to animals. For instance, it might be viewed as anthropomorphizing cows and calves when I recall the grief and panic they exhibit over their separation. There remains a widespread belief in academic and nonacademic circles that animals do not experience grief or loss or joy or love. And this is convenient in contexts where humans may cause bodily or emotional harm to animals: maintaining ignorance over the emotional lives of animals, and describing those who talk about animal emotion as anthropomorphizing or sentimentalizing, helps to maintain and excuse the very practices that cause animals trauma. If the cow is not traumatized by the loss of her calf because she can't feel trauma, then there is, perhaps, nothing objectionable about removing her calf hours after birth to divert the milk she produces to commodity production. However, if the cow and her calf *are* traumatized, and that trauma is recognized by human producers and consumers, then there arises a serious ethical dilemma of whether it is acceptable to knowingly cause that trauma, often repeatedly, over the life course of that cow.

Increasingly, this charge of anthropomorphism is being replaced with a less anthropocentric way of thinking about animals—that is, to say that animals have emotions is gradually being understood not as attributing *human* characteristics to animals, but instead *animal* emotions and their particularities are being better understood in their own contexts. Ethologist and evolutionary biologist Marc Bekoff argues that there is nothing necessarily wrong with anthropomorphizing; in fact, he argues that we must anthropomorphize. For Bekoff, anthropomorphism is using human terms to describe what we see in other animals; in other words, we use what descriptive tools are available to us as humans to understand nonhuman experience.[2] Bekoff, Jane Goodall, and Barbara King have written about the complex worlds of animals' emotional lives, highlighting how animals experience grief, love, joy, play, and fear.[3] But it is not necessary to be an ethologist to understand animal emotion. Careful attunement to animals' embodied states, and learning about how certain species and individuals tend to express themselves, is a window into knowing other species. This level of attention is something that can be developed through learning about what animal behaviorists and people working with particular species say, as well as through one's own observations and interactions.[4]

In the summer of 2014, I taught a class (called "Doing Multispecies Ethnography") on animals in the food system in which students were paired with a pig at Pigs Peace and were asked to write an ethnographic analysis—a sort of life history—of that pig. Over the term, the students observed and spent time with the pigs for several hours each week. At the end of the course, they all noted the uniqueness and singularity of each pig. They described in detail how smart or resourceful she was, what a great sense of humor she had, what her favorite foods were, how she interacted with the other pigs, and so on. During their final presentations of their work, we all realized that every single student had written about how exceptional their pig was, a revelation that revealed that pigs—and other farmed animals—are singular beings with distinct personalities and histories that make them each unique.

The students in the course also commented on how unusual it was to have contact in this way with farmed animals. And indeed, as urban areas boom and fewer people are living in rural areas and engaging in the practice of farming animals, contact with live cows, pigs, chickens, and other farmed species is increasingly limited. It is not merely geographic distance from farms, though, that enables human disconnection from getting to know animals like cows or pigs. It is also the way humans create hierarchies and categories of species. Humans are adept at categorizing particular species according to cultural norms: in the United States, for instance, dogs and cats as "pets"; rats, mice, and cockroaches as "pests"; cows, pigs, and chickens as "food." These categorizations maintain hierarchies in which humans are situated at the top, and they justify human use and treatment of other species in particular ways. That rats are "pests" or that the nutria is "invasive" in North America justifies their largely unquestioned eradication based on their species membership.[5] That cows are categorized as food— either as producers of milk or as beef—does important work to maintain their status as products for consumption. The hierarchical ordering of humans over animals is rarely conceptualized as problematic because of its normalization and the myriad ways in which humans benefit and profit from the appropriation of animals' lives and bodies. As Michael Parenti explains, "The most insidious oppressions are those that so insinuate themselves into the fabric of our lives and into the recesses of our minds that we don't even realize they are acting upon us."[6]

One example of this insinuation of hierarchy into daily life is the everyday language used to talk about animals. The term *cattle*, for instance, is not used in this book because it has its etymological roots in *chattel*, meaning property, and calls up references to chattel slavery. Instead, I use *bovine animals* or *cows* to signal that they are more than mere property. As an aside, *cows* is a colloquial term used to refer to bovine animals. In animal agriculture's technical language and binary way of thinking about sex, however, a cow is a female animal who has given birth to at least one calf. A heifer is a female who has not yet given birth. A calf is a male or female animal under six months of age (sometimes referred to as *bull calf* to indicate an intact male calf, for instance). A bull is an intact adult male, and a steer is a castrated male.

Livestock literally means "live stock" and reinforces animals' status as live property. The terms *farm animal*, *dairy cow*, and *veal calf*, too, each define animals in terms of their productive value to humans and reproduce the notion that these animals are bred specifically for this purpose. Instead, I use the terms *farmed animals* and *cows used for dairy* to reflect the fact that these animals are subjected to processes of farming and food production (and not that these inherently form their identity). When I do use *livestock* or *dairy cow* here, I place them in quotes to emphasize the contested nature of these terms. Finally, even the use of the terms *humans* and *animals* maintains a hierarchy—or rather, a binary, with humans on one side and all other animals on the other, despite the fact that humans themselves are animals. Language shapes how we think about and treat others. The language we use to talk about other animals helps to reproduce hierarchies of human exceptionalism, or it has the potential to engage in working to liberate other species from oppression.[7]

As these categories are maintained to enable our continued eradication or consumption of certain species, it can be much easier to just accept this ordering as the status quo and move on. It can be easier to believe the dominant ideology, as Melanie Joy writes in *Why We Love Dogs, Eat Pigs, and Wear Cows*, that eating animals is *normal, natural,* and *necessary*. If consumers and producers are committed to the status quo—to breeding, raising, slaughtering, and consuming animals for food—do they even really want to be connected to the individual, living animals who will become that food? Do they want to know them? Their favorite place to be scratched? Or their favorite treat?

For many, the answer is likely no. One of my students, after a couple of visits to Pigs Peace, declared that he had not been able to eat bacon or pork since his first day at the sanctuary. This is not an unusual response. When my partner, Eric, and I began raising chickens in our backyard before I started graduate school and fell in love with Charlotte and Emily (named for the Brontë sisters), we vowed never to eat another chicken. Getting to know a singular animal can cause a pro found disruption in how we think about and treat a particular species. It can disrupt the commodification process in which we treat animals as things to be bought, transformed into new things, and sold.[8] *Commodification* is a term used to describe the conversion of goods, services, skills, and resources into commodities. This concept is rooted in Marxist theory and refers to the process of taking something that did not previously have economic value and assigning it economic value. According to Marx, the commodity is the unit on which capitalism is built. Commodities are produced, sold, and bought to promote the circulation of capital in global political economies. Labor is an example of how the body can be commodified. For instance, animals labor in the production of commodities (milk, eggs, semen), but they are also themselves commodities, as when they are sold alive or dead (as com modity producers or as "meat"). Rosemary-Claire Collard and Jessica Dempsey call this a *lively commodity*, "live commodities whose capitalist value is derived from their *status as living beings*."[9] They theorize the lively commodity in the case of exotic pets and ecosystem services (a form of market-based environmental conservation), but the cow is a prime example of the lively commodity: she is not only a commodity herself (buyable and sellable as living capital), but she also brings new lively commodities into being, thus reproducing the commodity circuit.

There's little room for getting to know an animal's personality in the process of transforming them into commodities. Indeed, as I will explain later in the book, resistant personalities especially are bred out of breeds raised for dairy—and individuals whose dispositions "turn nasty" are routinely culled from the herd. The only personality permitted in commodifying dairy is the "happy cow," illustrated beautifully in the Real California Milk commercials that show humorous scenes with cows describing how happy they are to be in California. The tagline at the end of those commercials states: "Great milk comes from happy cows, and happy cows come from California."[10]

Coming face-to-face with a singular animal can disrupt routine ideas about the place of animals in society as well as the routine market activities in which consumers are involved. For instance, consumer practices might shift when the cow the consumer knows comes to mind when looking at a package of hamburger meat at the grocery store. This rupture in how people think about animals occurs when the animal is no longer relegated to "price per pound." Recalling the living animal can be a powerful way to connect "meat" as a product to the animal from whom it came.

Importantly, however, physical proximity to or visibility of farmed animals does not always foster a recognition of the animal as more than a mere investment or food source. For instance, farmers I met saw the individual animal insomuch as they were concerned with their health, condition, age, current productivity, future productive potential, and so on. These are all qualities and characteristics of singular animals to which farmers pay close attention. However, these features are all situated within an understanding of the animal as a commodity—as the potential price per head or per pound or as a prolific milk producer or not (as "milk" or "meat"). Conceptualizing the animal in this way obscures other yet more essential features of their lives. The identity of the animal is subordinated to market forces. It is not who the animal is but, instead, what—and how efficiently—she can produce.

The dairy industry itself—and the institution of animal agriculture at large—tends to talk about animals as abstract populations and not as embodied, singular beings. For instance, the terminology used in the industry has an abstracting effect: *cattle, beef, head (of cattle)*. In fact, some industry workers go so far as to refer to the live animals as *beef*. At one auction I attended, I heard an auction worker yell to another, referring to a group of "spent" cows destined for slaughter, "Hey, help me over here—let's move those beef out onto the truck." This abstraction from the living animal is at work in the practices of the industry, too. Efforts are made to standardize the productive capacities of the animal through breeding for consistency and through mechanizing production, in the case of milking machines, feeding practices, slaughter disassembly lines, and packaging. Animals receive brands and identifying numbers on ear tags to keep track of them in a herd, and they are rarely, if ever, named. Auctions sell cows individually or in herds, but the efficiency and monotony of the auction yard is such that the animals quickly blur into an anonymized stream passing through the ring.

The understanding of the animal-as-commodity that is born on the farm carries through to the site of consumption: the food purchase, the dinner table. Consumers are experts at engaging in the act of denial necessary to forget that the package of meat in the supermarket or the burger at the picnic table were living animals. Or they may not allow themselves to believe in the first place that "meat" is an animal. Jonathan Safran Foer, in *Eating Animals*, recalls his horror as a child when his babysitter reveals that the chicken he's eating is actually *a chicken*. Looking at a package of chicken breasts or ground beef, one does not have to confront the living animal—the "broiler" chicken confined in a shed with fifty thousand other birds, the "spent dairy cow" on her way to slaughter. It is even easier to deny the animal involved in a product like milk, where one is not confronted so directly with the animal (their dead body) at the moment of consumption. With a commodified mammary excretion like milk, it is possible to imagine a scenario where the animal is not harmed for the production of that product.

One of my primary motivations for doing this research and writing this book was to understand the way dairy is produced. Contrary to the way many people now understand meat production as involving harm to the animal, public perceptions maintain that the production of dairy is benign. I wanted to know the details of where milk, as a sellable good, comes from, how it is produced, and with what costs to the lives and labors of other species. Many studies of milk and dairy thus far focus on milk as either a cultural artifact, dedicated to understanding its meaning in cultural histories around the world, *or* as a product at the heart of a primary economy of food production, emphasizing the impacts on dairy farmers, falling milk prices, and consolidation in the industry.[11] What I have written here is different. My aim from the outset of my research was to understand the impacts of commodification on bovine animals themselves in the dairy industry—the cows, bulls, and calves—who labor to produce milk for sale and human consumption. Thus, farmers and industry workers are present in the book, but the focus here is less on them and more on the cows they farm. This focus on cows is not to suggest that the lived experiences of farmers and farmworkers is unimportant. Indeed, farmers struggle to get by; falling prices for agricultural goods and the increasing consolidation of agricultural industries drive many farmers off land their families have farmed for generations. Wage farmworkers are a highly exploited workforce, as they engage in dangerous, difficult, and underpaid work with little recourse for labor organizing or demanding fairer working conditions. Depression and suicide occur at alarmingly high rates among people in farming professions, and a US Center for Disease Control and Prevention study published in 2016 reported that those in farming, fishing, and forestry occupations commit suicide at significantly higher rates than those in other occupations.[12] Wendell Berry has written eloquently for decades about the plight of farmers and farming families in the United States. And there are many excellent books that outline the violence and exploitation experienced by farmworkers in the United States (Timothy Pachirat's book *Every Twelve Seconds*, Steve Striffler's *Chicken*, and Seth Holmes's *Fresh Fruit, Broken Bodies*, to name a few). Without discounting the very real human struggles and suffering that occur in the production of food, then, I have dedicated the pages in this book to focusing on other animals' experiences of food production.

Why do many consumers in the United States and beyond shy away from knowing how dairy, eggs, or meat are produced? To some extent, most people seem to already know, and that is why they turn away. At social gatherings, when people ask me what I do for a living, I reply that I teach and I study the lives of animals in the food system. More often than not, I am abruptly interrupted with: "Don't tell me, I don't want to know!" or "Oh! That must be disturbing!" So to some extent, many people do already know. Or they know enough to know that they don't want to know more. There is an inkling that raising animals for food involves a certain level of violence or at least a certain amount of something they don't want to think about. It is not uncommon in today's proliferation of social media to have seen an undercover video or two of animals suffering in the food system or, perhaps, to have heard about what these films depict.

Most people in the United States say that they do not want animals to suffer. A 2013 consumer survey, conducted by the American Humane Association, found that 89 percent of consumers were "very concerned about farm animal welfare," and 74 percent said they would pay more for products that were labeled as "humanely raised."[13] An earlier survey, conducted by Oklahoma State University in 2007, revealed that most respondents believed that they care more about animal welfare than the "average American."[14] For example, 95 percent of respondents agreed that it was important to them how farmed animals were treated, but only 52 percent believed that others felt the same way. Similarly, this survey found that 76 percent cared more about farm animal welfare than about low meat prices, but only 24 percent believed that others shared this opinion. These results suggest, according to the Oklahoma State researchers, that, in survey responses, consumers maintain a flattering image of themselves, imagining that they care more for animal welfare than others, even if this stated concern does not translate into changes in behavior or consumption practices.

Whether people do actually care about the welfare of farmed animals or whether they may like being seen to care, the reality remains that most people are adept at forgetting this concern when they sit down to eat. Whether it is convenience, routine, tradition, family or peer pressure, or forgetfulness, concern for farmed animals often falls to the wayside in daily practice. As Foer describes, even those "selective omnivores" who vow to buy animal products only from small-scale farms or farms that claim to treat their animals better will frequently give in to pressures of convenience or social grace when it comes time to eat.[15] Alternatively, there are many people who believe they do not eat much meat or dairy at all, but who, in fact, consume animal products with nearly every meal. These discrepancies in how people believe they behave and how they actually behave are enabled by the ease of forgetting the animal. It is so easy to forget or deny the things that are reminders of the animals at the start of the food commodity chain. So easy to never see them or think of them in the first place.

It is for this reason—to combat the act of forgetting and looking away—that I have written this book. Why is this book about dairy and not about meat production or some other industry of animal use? As I stated earlier, one of my motivations for researching and writing about dairy is the public perception of dairy production as benign. This perception is complex and is tied up with marketing and advertising by dairy industry lobbies (e.g., the Milk: It Does a Body Good and the Got Milk? campaigns). Human milk is also our first food, and so cow's milk becomes an extension of the way we are first nourished. Dairy products are also rich with nostalgia as part of a US imaginary: ice cream cones dripping in the hot summer sun, gooey macaroni and cheese, grilled cheese sandwiches and tomato soup, a tall glass of milk with cookies. These classics in the mainstream US foodscape are woven into memories of childhood and eating for comfort and pleasure. These modes of nostalgic longing also help facilitate the denial or intentional ignorance about how these foods are produced.

Another important dimension of research on the dairy industry involves the fact that the process of producing dairy is not just about extracting milk from a cow. The production of milk is tied directly to a variety of other industries and practices we may not think of when nibbling on that slice of cheese: veal and beef, artificial insemination, forced semen extraction, animal feed industries, rendering, and other peripherally related industries. The details of how dairy is produced (and everything implicated in that production) are obscured from public knowledge. Many well-educated and thoughtful people I've talked to throughout my research process, for instance, were surprised to discover that a cow has to be regularly impregnated to produce milk. Of course, they do. Like any mammal—humans included—cows produce milk after they have recently given birth in order to provide much-needed nourishment to their young. But this obvious point is obscured in the magical thinking—or more likely, the lack of thinking, with which people imagine dairy production.

Dairy production practices are also sustained by long-held ideas about the necessity of consuming milk for human health. Élise Desaulniers's book *Cash Cow: 10 Myths about the Dairy Industry* challenges dominant narratives about the healthfulness of dairy, offering resources for recent research that debunk norms about the necessity of dairy as part of the human diet. Milk Truth is a dairy industry campaign designed to "tell the truth" about milk; focused explicitly on the healthfulness of milk, it offers reassurance that consuming milk is not only healthy for humans but also necessary. Milk Truth's website reads: "For some reason, milk has been under attack. Some critics are saying don't drink milk—it's unneeded, unnatural, and bad for you. That couldn't be further from the truth. Thousands of scientific studies have documented the benefits of drinking milk. Don't be misled by alarming headlines or passionate critics. Get the full story about milk. Nutrition is a science, not a point of view. See what the real experts are saying about milk—one of the most naturally nutrient-rich beverages you can find."[16] Desaulniers confronts these claims in depth in her book, putting forth an alternative approach to thinking about dairy industry discourse and research. Milk Truth asserts that they are providing "The Milk. The Whole Milk. Nothing but the Milk." Both dairy industry proponents (e.g., Milk Truth) and critics (e.g., Desaulniers) present their research and claims as the truth   a particularly interesting formulation during the Trump Administration's age of "alternative facts." Milk Truth frames their perspective as being objective truth in opposition to the "passionate critics" of the industry and they claim that "nutrition is a science, not a point of view," implying that their "truth" is indisputable and not shaped by social, political, or emotional (*passionate* can be read as code for *emotional* or *sentimental*) influences.

Nutrition is, indeed, a science, but how nutritional recommendations are arrived at and made is a highly politicized process that is deeply entangled with industry interests, lobbying, and complex political negotiations. This process, in the context of US federal nutrition guidelines, is documented extensively by Marion Nestle in her book *Food Politics*, which traces the history of nutritional and dietary guidelines in the United States and the political power and fraught negotiations that constitute these guidelines. Nestle outlines how, in 1977, for instance, the US Department of Agriculture (USDA) drafted a report titled *Dietary Goals for the United States* that recommended that people eat less meat, eggs, and high-fat dairy products; the outrage registered by the meat and dairy industries was significant, as they feared that this would dramatically cut into sales of the commodities that they produced, and the committee eventually released a revised, more watered-down report that avoided "eat less" kinds of language and favored language like "eat more lean meat."[17] Over the past few decades, the meat industry has lost some ground in their representation in the USDA's dietary guidelines; the current MyPlate guidelines, produced by the USDA Center for Nutrition Policy and Promotion, for instance, includes a "protein group instead of a meat group, reflecting an inclusion of the much broader range of protein-rich foods that comprise US diets (including beans, nuts, tofu, and vegetable-based meat replacers).[18] The dairy industry, however, has been more successful in lobbying to maintain strong representation in the current dietary guidelines with its own dairy category prominently displayed just next to the plate. Thus, ideas about what's healthy or necessary (and how milk and dairy products factor into US diets) are a deeply fraught, political process that is obscured by the dairy industry's assertions that "nutrition is a science, not a point of view."[19] But, in fact, scholarship like Nestle's meticulously details how nutrition science does reflect a point of view, as well as deeply rooted political and economic agendas. With this brief history of nutrition science in mind, let me now turn to some background on dairy production in the United States.

DAIRY FARMING IN THE UNITED STATES

At the end of 2016, there were over 9.3 million cows in the US dairy herd.[20] Dairy production is tied closely with the meat industry, which slaughtered 2.93 million "dairy cows" in 2016 for (primarily ground) beef and 501,500 calves for veal during that same period.[21] These statistics are important for understanding the scale of production in the US dairy industry, but they have the potential to obscure the lived reality of each of the 9.3 million cows living and laboring for dairy production each year. After all, these numbers can have an abstracting effect. As David Wolfson writes, "When discussing the treatment of such a large number of animals, it is hard not to write either in a droning monotone or somewhat sensationally. . . . It is not simply more than [9.3 million] animals a year, but it is one, and one, and one, amounting to the large scale mistreatment of individual animals."[22] My aim here is to make legible the stories of the manifold "one"—the singular animals in the dairy industry—and, without sensationalizing, to take their experience seriously. At the same time, these stories can say something meaningful about the plight of animals raised for food more generally.

The lives of animals in the US dairy industry are characterized by intensive management and manipulation of the reproductive and productive capacities of their bodies. The use of bovine bodies for breeding, milk, semen, and meat is highly gendered, based on the industry's assessment of the animal's biological sex at birth and notions of biological sex as it is tied to reproduction. In so doing, the dairy industry reproduces binary ways of thinking about sex and reproductive capacity and reinforces gendered ideas about the body. Animals are framed in the industry, first, as being "female" or "male" at birth and, then, as being reproductively viable or not; the trajectory of their lives is organized around these two logics.

Cows are used for the production of milk until their productivity wanes and they are slaughtered at three to seven years of age, while most male calves are slaughtered at approximately four to six months of age for veal. By contrast, the natural lifespan of a dairy-breed cow is upwards of twenty years (see fig. 1.2). This connection between dairy and slaughter is one that is under-recognized in public consciousness, just as the many facets of routine dairy production—artificial insemination, semen production, feeding, tail docking, castration, dehorn-

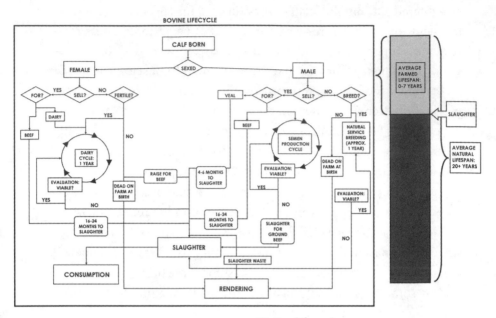

FIGURE 1.2 Bovine life cycle

ing, birthing, milking, transport, sale, slaughter, and rendering—are
largely absent from the popular image of dairy production.

The remarkable consolidation and intensification of the dairy industry has very real impacts on the lives of cows laboring on farms
throughout the United States. From 1970 to 2006, the number of
dairy farms in the United States decreased by 88 percent, the number
of cows in the US herd fell from 12 million in 1970 to 9.3 million in
2015, yet simultaneously, milk production doubled.[23] More recently,
between 2001 and 2009, the number of commercial operations with
milk-producing cows declined by 33 percent yet milk production increased by 15 percent while the number of cows in the US dairy herd
remained relatively steady with an increase of 1 percent in that same
period.[24] Meanwhile, USDA records show that between 1970 and 2006
the number of large-scale farms (those housing five hundred or more
cows) increased by 20 percent and, in particular, those with two thousand or more cows increased dramatically—by 128 percent.[25] Some
large dairies in the United States house well over fifteen thousand

cows, while the more common large-scale dairies house between one and five thousand cows.[26] The number of small farms (with fewer than five hundred cows) declined by 35 percent between 2001 and 2009, and the number of the smallest farms (with fewer than thirty cows) produced just 1 percent of all milk, even though they accounted for nearly 30 percent of the total number of US dairy farms.[27] This trend indicates consolidation in the industry—meaning that fewer cows are producing more milk on fewer farms than ever before—as small farms close in favor of more industrial-scale production. And this reflects a process familiar throughout much of modern agriculture—the consolidation and intensification of food production—which magnifies the impact on humans and nonhuman animals.[28]

Geographically, dairy production is growing in the western United States and large-scale dairies in particular are increasing in western states. California is by far the most prolific state for dairy production, accounting for 19.6 percent of all US dairy production.[29] Seventy-four percent of all US dairy production in 2015 occurred in the top ten dairy-producing states (in descending order of percentages of overall production): California (19.6 percent), Wisconsin (13.9 percent), Idaho (6.7 percent), New York (6.7 percent), Pennsylvania (5.1 percent), Texas (4.9 percent), Michigan (4.9 percent), Minnesota (4.5 percent), Missouri (4.2 percent), and New Mexico (3.7 percent).[30]

I chose to focus my research in California and Washington—both of which are in the western United States—to reflect the geographical shift of dairy production westward. As the largest dairy-producing state in the country, California was an obvious choice. California's Central Valley features many industrial-scale dairies, some with upwards of fifteen thousand animals per farm and approximately one-fifth of national dairy is produced in the state. I chose Washington for different reasons. While Washington was in the top ten of dairy-producing states when I began my research (as of 2015, it was the eleventh top dairy-producing state in the country), it accounts for only 3.1 percent of national production.[31] I chose Washington because it is the state in which I reside, and while this may seem to be a trivial reason, my feminist geographic training makes me attuned to the way the places we live and work are important sites for deep intellectual and ethical engagement. Washington is also an interesting case for its unique geographical-agricultural climate around the Puget Sound region, featuring smaller-scale farms serving niche populations interested in supporting local, organic, and small-scale farmers. Looking at these two scales against one another is important in order to draw out the nuances of each scale of production. Studying only industrial-scale production, and making an argument about dairy production in general based on this scale alone, for instance, erases the particularities of important scalar differences, as well as the opportunity to report the (at times, striking) similarities between the two.

My research for this book involved two years of in-depth content and discourse analyses of web and print-industry materials, such as advertisements, how-to manuals, and industry publications; university agricultural extension manuals used in educating the next generation of dairy farmers in the latest practices and technologies involved in milk production; scholarly publications on animal agriculture, behavior, and welfare; and US state and federal law governing animals in spaces of agriculture. To augment this analysis, my research also took me to various places of dairy production and other places where the lives of animals used for dairy could be understood. I traveled to a mid-sized dairy farm in western Washington and to sanctuaries for formerly farmed animals in California, and interviewed the people who worked in these spaces. I attended, as a spectator, three different auction yards in California and Washington on multiple occasions, the World Dairy Expo in Madison, Wisconsin, and the Washington State Fair in Puyallup, Washington. I explored peripheral but integral industries related to dairy and became familiar with the details of semen production, veal production, and the rendering process. I drew on previous research I had done on the slaughter process in the United States at industrial and small scales. I interviewed people who were involved with 4-H (a global consortium of organizations for children's education and development, with a focus on farming in many contexts) as children and former animal science students from university agricultural animal science programs to explore the role of education in the reproduction of the dairy industry. Each of these places inhabited or touched (even for a moment) by the living or dead cow revealed something important about the place of animals in the dairy industry.[32]

I focused on "best practices" within the industry: the laws and guidelines that define how raising, handling, and slaughtering animals *should* be done. The farm I visited in western Washington was on the smaller scale, where those working with the cows took great care in their daily encounters with them. The auction yards were mundane, everyday spaces of exchange. There is something particularly interesting about analyzing best practices in animal agriculture and slaughter. It is easy to criticize practices that clearly violate legal guidelines for raising, handling, and slaughtering animals; incidents categorized as animal abuse and cruelty are, as one might expect, denounced within the industry and among government overseers of farming practices. There is widespread consensus that actions constituting cruelty or abuse should be reduced and eliminated in animal agriculture. What's more interesting, and what became the focus of my research in this book, is to look carefully at what happens in the dairy industry on a *routine* day. What was illuminated for me at the farm, at the auction yards, in documentation of slaughter practices, and in other industry spaces was that routine, USDA-approved industry practices—the way things operate on their *best days*—still enact violence on animals raised for food.

It may seem hyperbolic to call routine industry practices *violent*. Indeed, through my research, writing, giving talks, and engaging in conversations with people about this work, I have encountered regular objections to using a term like *violence* to describe what I detail in this book (and these objections have even come from well-established animal studies scholars). But this, in itself, reveals important insights about how animals raised for food are cared about and conceptualized: how their lives and deaths come to matter (or not) in ethically fraught ways. What I've become most interested in, and what I hope to illuminate in this book, is the way violence against certain lives and bodies can become so normalized that it is not viewed as violence at all. And this is relevant not only for thinking about violence against nonhuman animals but also for thinking about how violence, more generally, is sustained, reproduced, and erased by structural mechanisms (like law and capitalism), by social norms (like histories, cultural practices, and dominant discourses), and by frameworks of inequality (like human exceptionalism).

WAYS OF KNOWING OTHER SPECIES?

The nature of dairy and related industries and the limited access to places of animal agriculture explored in subsequent chapters are such that full life histories of animals were nearly impossible for me to obtain. Transport, sale, slaughter, and the general culture of commodifying cows in the industry limited what was knowable about each animal. One of the central foundations for this work is the question of how to begin to know the experience of a member of another species. This is an especially complicated question when the species in question is thoroughly entangled in institutions of domestication and agricultural commodity production. How to know the impacts of being transformed into a commodity on the life of the cow is a fraught question, dependent not only on the singular animal and social relationships on the farm but also on the positionality of humans who encounter her: farmers, auction workers, slaughterhouse employees, sanctuary caregivers, researchers. The cow's framing as a domesticated, commodified being shapes how knowledge is produced, circulated, and contested about her, and about cows more generally. The abstracting logic of commodification works against the intimate knowledge of the singular cow's inner world and life experiences. Thus, ways of studying, knowing, and encountering cows are necessarily a politicized process. Viewing the cow not merely as a commodity (to be managed, made more efficient) but also as a commodified being whose life is wholly shaped and determined by her commodification is a political statement on its own.

Of course, farmers and industry workers involved in the daily care of animals have much more sustained contact with animals on farms than I did as a researcher. Although, increasingly, even farmer contact with cows raised for dairy is limited, as the number of cows on each farm increase (and individualized attention becomes less common), and as processes like transport and slaughter often occur outside of the farmers' control or oversight. Even so, and this is a central insight for my approach in this book, farmers and industry workers know the animals they work with in a *particular kind of way*. They know them within the confines of their conceptualization as commodities. In other words, they might know that cow #4789 needs a vaccine or will be giving birth soon or that she has maybe one more round of birthing before they will send her to slaughter. They will know that this particular cow was born of this particular sire who is well-known for breeding cows with high-quality milk. But these are ways of knowing that situate the cow's singularity in a logic of commodification and that can obscure the fundamental impacts of this logic on her life and being.

Importantly, these are ways of knowing that *depend* on *not knowing* other things that would complicate the farmer's role in the process of transforming the cow into a marketable good. These ways of knowing depend on not knowing (or acknowledging) things like the fact that cow #4789 might be experiencing profound grief at the removal of her calves each time she gives birth, or that she might prefer not to be slaughtered for meat. Commodifying farmed animals is made easier—indeed, is enabled by—not thinking about the emotional and psychological impacts of this process. And to be clear, this is not meant to vilify farmers or industry workers in any way. Consumer demand for animal-based commodities requires farmers and industry workers to meet this demand, and so not thinking about the cow's emotional distress may be an emotional survival mechanism for the farmers and workers themselves, or it may be that the pressures of work and increasing speeds of production simply don't allow time for this kind of reflection. And more expansive and pervasive than consumer demand and its implicatedness in this process are the broader economic logics that perpetuate how humans commodify living beings in the first place.

To be certain, I have not spent nearly as many hours or days or years with cows as farmers have. However, throughout my research, I realized that my outsider status has allowed me to see and understand particular dimensions of cows' lives not readily visible to, or acknowledged by, farmers or industry workers. Feminist theorist Donna Haraway's theory of "situated knowledges" critiques the notion of objective knowledge, arguing instead that *all* knowledge is partial, arising from particular positions and embodied experiences.[33] Thus, my research on the lives of cows is built from my own positioning, my own way of knowing, that is a unique constellation of my academic training, my personal history and experience, and my situatedness outside of farming communities. And, of course, I will not know or be attentive to certain things as a result of this situatedness. I am not as attuned to the subtle nuances of the contours of a cow's ribs and what that signals about her meat-producing potential as a meat buyer in a cull market auction would be. Or I don't readily know a cow's worth in dollars by looking at her, and whether the seller has priced her fairly.

Less constrained by the constricting logic of the animal-as-commodity, and having carefully reviewed animal behavior scholarship on bovine behavior, stress, and welfare, I was attuned to different things—things unnoticed, or maybe noticed and ignored, by the meat buyers in the room: the emotional trauma of the cow separated from her calf in the auction ring or the labored breathing of a cow collapsed in the auction pen as she lay there for hours, unable to rise, or the repeated vocalizations of cows packed tightly in the holding pens, their eyes rolled back into their heads and mouths foaming with saliva. Informed by this scientific scholarship and by my own observation, my ongoing instruction of a course on multispecies ethnography, and my involvement with sanctuary work, my positionality offers a radically different way of knowing these individual animals. Importantly, this way of knowing also involves an empathetic approach. I have written elsewhere that fruitful ways of knowing animals in the dairy industry can be built from engaging what Lori Gruen calls "entangled empathy" in the research process, a process in which I acknowledge and center the ways in which humans and animals are already always caught up in relationships of power, control, and care.[34] Empathetic engagement helps to highlight what these relationships tell us about their effects on a *particular life*. This kind of engagement—paired with knowledge informed by studies of animal behavior and emotion—can provide robust counter-knowledges to the dominant way of knowing farmed animals promoted by animal-use industries.

To illustrate this point about situated knowledge of farmed animals, I share an anecdote from Judy Woods, the director of Pigs Peace Sanctuary. Judy had enormous difficulty, when she started the sanctuary in the 1990s, finding a vet knowledgeable about pigs. Judy knew nothing about pigs when she founded the sanctuary; everything she knows now, she learned from working closely with the pigs. She began by reading everything she could find about pig care. Everything she found was written by farmers or published by agricultural extension programs, and the more she lived and worked closely with pigs the more she realized that common and published knowledge about pigs, their care, and how they prefer to live went completely against what she was seeing in the pigs who were in her care. As she struggled to find a vet who could provide the kind of medical attention the sanctuary pigs needed, she encountered a disjuncture in dominant ways of knowing pigs (coming out of agricultural communities) and her own experience learning about pigs (in the context of the sanctuary). Because pigs are conceptualized as commodities and framed within rational calculations of profitability and its effects on care, pigs are often given little or no medical treatment. And because pigs are conceptualized as commodities, aging pigs are an anomaly on farms; farms do not tend to house geriatric pigs who will need specialized veterinary care and attention as they age and their bodies break down because they are typically slaughtered long before they experience any effects of the aging process.

Contrary to the way vets in the United States regularly see geriatric cats and dogs who suffer from arthritis, cancers, loss of sight or hearing, and other conditions that occur with aging, vets who treat farmed animals focus their attention and education on learning other features of care: how to help a pig stay healthy enough to reproduce until her fertility declines and she is slaughtered or how to treat pigs with antibiotics prophylactically to stave off disease and infection in

highly crowded feeding operations. In spaces like Pigs Peace Sanctuary, where animals are not defined by their commodity potential, the
veterinary needs—the *ways of knowing* pigs—are very different than
on the farm. Over the years, Judy worked with a vet trained in large
animal care, and shared her own situated knowledge with him to develop a more holistic program of care for the pigs at the sanctuary. This
shared knowledge production (his formal veterinary training, which
lacked certain important features of caring for pigs who are not raised
as commodities, and her intimate knowledge drawn from daily care
and interaction with the pigs themselves) combined to transform the
possibilities of medical care and ways of knowing pigs in this context.

Central to this book is the point that the act of commodifying nonhuman animals like cows shapes how humans know them, how humans care for them in different spaces, and how knowledge is produced about them. While this book is first and foremost about cows
and dairy production, it is evident from my weaving in of experiences
and reflections not related to cows (e.g., the inclusion of ruminations
on Pigs Peace) that it is also, fundamentally, about ways of knowing
and being in relationships of care and harm with other species. Indeed,
the knowledge collected in this book reflects the nature of commodification: the effect is highly fractured, often momentary, and woven
together in a sort of patchwork of research and life experiences with
cows and other animals. Many of the stories I share are mere moments
in the animals' lives, fragments of much more complex commodified
lives, told from my perspective as a researcher and outsider in spaces of
farming. The cow with ear tag #1389, for whom this book is titled, oc-
cupies only a few pages in this text. The poignancy of our momentary
meeting and its fleeting nature acts as an embodied reminder of what
is entailed in raising animals for capital accumulation. Thus, while
this book focuses on encounters with a number of singular, individual animals, my passing encounters with these animals highlight the
fractured quality of their lives and social relations and the abstracting
effects of commodification from the lived experiences of those who
are commodified.

GRIEVING SADIE

It was June of 2012 when I met Sadie and walked around Animal Place with Marji Beach, learning about the animals who live there. In August of that same summer, Sadie died at the sanctuary. When she died, tens of thousands of people around the world grieved. All those who followed the work of the sanctuary through newsletters and social media updates were familiar with Sadie's story and the way she made her way to Animal Place. This level of grief for a single cow is certainly extraordinary. When most farmed animals die, their deaths go unremarked on by humans. Most of the animals whose stories fill the pages of this book will not be (or were not) mourned at their deaths by humans (although they might certainly be mourned by other animals with whom they shared relationships). This is just one way that Sadie's story is exceptional. The fact that she made it to sanctuary made it possible for her be mourned. But Sadie was also, for years before her move to the university and eventually to sanctuary, an ordinary cow in the dairy industry. Her life at the dairy farm, her presence at auction, her chronic case of mastitis, and the other physical traces of dairy production left on her body represent the mundane reality of cows raised for dairy in the United States.

2

THE POLITICS
OF RESEARCH

One of the most surprising discoveries I made during my research was the repeated rejection I received from all but one dairy farm I asked to visit. I was ready to be turned away from large-scale producers since I had heard that industrial-scale farms are almost always closed to outsiders (journalists, the public, researchers). What I wasn't ready for was the rejection, the suspicion, and—in some cases—the hostility of many of the small-scale dairy farmers I contacted. The process of doing research—and academic research, in particular—about spaces of food production is highly political, and fear and secrecy perpetuate a culture of silence around spaces of food production. The changing landscape of access to spaces of food production and the growing lack of transparency marks a shift in the political climate of food and animal-use industries and bears outlining here.

When I set out to do this project, I planned to apply for a job on a large-scale dairy farm. Because it is increasingly difficult to gain access to large-scale farms as a researcher or journalist, it is not uncommon for those writing about animal agriculture to apply for work without disclosing their position as a researcher to gain access to these spaces. Other scholars interested in food production, industrialization, and labor in animal agriculture have successfully found work on farms and in slaughterhouses in the United States to complete in-depth projects on these places. Timothy Pachirat, for instance, worked for five months in a Nebraska slaughterhouse to write an analysis of slaughterhouse labor for his doctoral dissertation, which later was published as the book *Every Twelve Seconds: Industrialized Slaughter and the Politics of Sight.* Prior to Pachirat's book, Steve Striffler worked in a slaughter plant for chickens for his book *Chicken: The Dangerous Transformation of America's Favorite Food.* Informed by this established tradition of research, I began to design the methods I would use to learn about dairy-farming practices in the United States.

University research involving direct contact with human beings is required to go through the Human Subjects Division of the ethics review board, called the Institutional Review Board (or IRB). This process is meant to ensure that researchers who make contact with and study humans are doing so in an ethical manner. In other words, IRB approval involves demonstrating how your research design takes into account the need to protect the identity and privacy of research participants as well as their safety and well-being throughout the course of the study. If deception is required for the study (withholding information about the nature of the study, the role of the participants, or any other information), researchers are required to make a compelling case and justify this deception in their application.

Based on my initial contacts with regional dairy farms in the western United States, I discovered quickly that farms were unwilling to have a researcher visit and study the farm. Obtaining undercover employment in an industrial dairy farm, then, was my next best option, but this would have required that I withhold information in the job application process about my identity as a researcher. This meant that my project involved deception. I made a case with the IRB for why this deception was necessary and how I would mitigate any potentially negative impacts, and I drew special attention to the fact that this was an established tradition among researchers of animal agriculture and other industries not widely open to the public.

Immediately, I was told by an IRB staff person over the phone that the university would not approve this use of deception. She said that deception was reserved almost exclusively for medical studies where subjects were not informed about whether they had been given a placebo or an active drug, for instance. She reported that the form of deception that I was proposing would open the university up to serious liability issues. She said that my IRB application would be approved only if I proposed disclosing my identity as a researcher up front to all farms and potential interviewees I contacted.

My ethics review approval process was further complicated by the inclusion of nonhuman animals in my research. During the review process, researchers are asked whether their research involves the use of nonhuman animals as research subjects. I answered yes in response to this question, which caused my application to be routed through the health sciences review process, rather than the social sciences, since the university overwhelmingly conceptualizes animals used in research as those animals who are used in *biomedical* research in labs. Answering yes to this question also meant that I had to complete the ethics review process for nonhuman animals through the Institutional Animal Care and Use Committee (or IACUC), a division of the Office of Animal Welfare. Although animals as research subjects are becoming increasingly common in field research across the social sciences, institutional programs are designed to evaluate the use of animals in biomedical research and not in qualitative social science research.

To give an example, in order to obtain IACUC approval for my project, I had to complete an online training module for research involving animals. The training course was disproportionately focused on approved methods of euthanasia for animals in laboratory research, dedicated primarily to ensuring that animals are killed by one of several approved methods depending on which species they belong to. So, for instance, through this training, I learned that decapitation is an approved method of killing a young rat at the end of a study, and I learned "best practices" for rat decapitation, but I learned nothing about ethical research practices involving cows or other farmed animals in qualitative social science fieldwork.

During my mandatory meeting with IACUC personnel about my project, I was asked about the nature of my interactions with the cows I would study.

I said, "I'm going to observe them, and I might be involved with feeding and watering them."

"What else?" the staff member asked.

"Um, well, I guess I'll probably pet them!" I added, racking my brain for what she was looking for, trying to be helpful. She looked at me blankly and then burst out laughing.

She informed me that the IACUC did not have specific protocols for animals on farms and that, as a result, I would merely need to adhere to any federal and state laws governing farmed animal welfare. I informed her that the only legislation that protects farmed animals in the United States is the Humane Methods of Slaughter Act, which governs the slaughter process, but not how the animals are raised prior to slaughter.

"Make sure you follow that, then," she said.

"I don't plan to be involved in any form of slaughter," I replied.

And that was that. In other words, the IACUC had almost no infrastructure set up for overseeing the research of animals on farms—or defining what ethical research practice might look like when the research subject is a farmed animal. If I had said that I would have been involved with killing the cows, they would have approved it so long as I followed the federal "humane slaughter" guidelines. The very concept of *ethical research* in contexts of human versus nonhuman animal studies highlights the unevenness and the entrenched anthropocentrism of the university as a knowledge-making institution. Calculations of killing and harm are regulated in only very limited ways for nonhuman animals involved in research (e.g., animals in labs, and approved methods of killing). That animals can be killed at all for research is a symptom of the hierarchical ordering of life and the disposability of individuals of certain species. I quickly realized that I would need to define for myself more stringent guidelines for ethical practice with the animals I encountered. But this, itself, is problematic—that a researcher would be in a position of determining for herself what constitutes ethical research because the institutional process did not suffice. I thought often about this issue throughout much of the research process and in subsequent years as I have undertaken new research involving animals.

If the motivations for my study were to understand the impacts of dairy production on the animals in the industry, how could I do the research and, at the same time, be sure that I was acting in an ethical way toward the animals I was researching? I was already familiar with undercover footage put out by animal rights groups of animals in the dairy industry, and I knew that widespread mistreatment of animals did exist in the industry. I had seen videos of living but immobile cows shocked repeatedly with electric prods and pushed with tractors. I had seen film of the blinking eye of a cow, left for dead on the "dead pile." I had seen a recording of a tiny newborn calf, still wet with amniotic fluid, tossed like a sack of potatoes into a wheelbarrow and wheeled away from the cow who bore him. What if I encountered one of these cases? What was I to do? What if I came face-to-face with one of these animals? How would I respond? As a researcher, was I expected to maintain an emotional detachment from what I might see? Going into the research, I imagined that my impulse in a moment of see- ing an animal suffering would be to *do* something—but what would (or even, practically speaking, *could*) I do? And what would this mean for my status as an observer? To my surprise, I found—much later in the research process—that I was able to watch these moments in my research and do nothing, and I am ashamed by this fact.

I worried about these issues as I redesigned my project in response to the IRB's rejection of my original plan to use any deception. In the revisions, I proposed that I would contact farms, declare my position as a researcher, and ask if I could either volunteer my labor or, at the very least, visit the farm and learn about their dairy production prac- tices. In addition to interviews with farmers, I would use participant observation—a method where I would observe the practices of partic- ular spaces through participation in various public events: the World Dairy Expo, the Washington State Fair, or farmed animal auctions, for example. I was already in the midst of extensive textual analysis, which formed the core of my research. This IRB- and IACUC-approved field- work would supplement what amounted to two years of textual re- search.

After four long months of back and forth with the ethics review boards, I received the necessary approval and I was finally able to begin my field research. At this point, I had completed not only significant textual research but also legal and video research on dairy farming. I decided I was ready to visit farms, and so I began contacting farmers in the spring of 2012. I approached both larger-scale, industrial farms as well as small-scale farms that sell their products at farmers' markets and local shops. I visited farmers' markets and talked with sellers of dairy there—they seemed excited and enthusiastic about my project and the possibility of my visiting the farms to do research. They encouraged me to follow up via phone or email to set up a time to visit. Of the many farms I contacted, I got a direct response from nearly two dozen farms. Of these, only one allowed me to visit. Others gave various reasons for not allowing me to come to the farm: time constraints, disinterest in my project, and biosecurity. To the time constraint reason, I asked if it might be better for me to come in a different season—I told the farmers that I would be working on this project for the next several years, and I was available at any time in any season to visit the farm—even just for an hour. I was told that there would never be a good time.

Many of these small farms claimed on their websites—and at their farm stands—that they raised their animals under the best conditions. "But don't take our word for it," they urged, "come and visit and see for yourself!" For this reason, I was especially surprised when I was turned away each time I tried to schedule a visit. I expected this level of secrecy and unwillingness to be transparent from larger, industrial-scale farms—industrial-scale producers have long been wary of opening their doors to people interested in the production process. What I was not expecting was this lack of transparency and the shroud of secrecy around small-scale, local production practices. However, this culture of guardedness on the part of small farms signals a more general shift in the political climate around food production and consumer knowledge in the United States: a shift whereby smaller farms are taking on the practices, technologies, and approaches to farming animals that are common in industrial-scale facilities. And my sense is that this is certainly not done with any malevolence; in an increasingly consolidating and competitive industry with close profit margins for dairy, small farms do what they must to survive.

One feature of this increasing shift toward more industrial approaches to farming (even for those operating on a small scale) emerged in the use of *biosecurity* as a discourse deployed by small farms to justify their closed doors. Indeed, biosecurity was the most interesting reason I encountered for not being allowed to visit—both because of its link to large-scale production practices and because of the ambiguity of the term and how it is used. In the dairy-farming context, *biosecurity* refers to a growing fear of contamination from outside pathogens, a threat exacerbated by increasingly intensive production practices and the division of the industry. For instance, higher concentrations of animals in smaller spaces increase the chances of disease vulnerability and risk of its spread among the population. And the separation of calves from the cows only a few hours after birth increases the calf's vulnerability to disease that would otherwise be limited by his nursing directly from the cow—getting the necessary antibodies from the colostrum that all young mammals get from the adult's milk.

Biosecurity is, no doubt, a viable concern for increasingly industrializing farms; however, the sense I got from these conversations was that biosecurity was more regularly used as an excuse than as a legitimate concern. For instance, many workers on dairy farms arrive to work and complete their jobs with no protective gear or precautionary measures to prevent the spread of disease to the animals. Many animals are treated prophylactically with antibiotics to prevent the contraction and transmission of disease and to promote growth. In fact, the Food and Drug Administration estimates that far more antibiotics are used by farmed animals than by humans.[1] And the Centers for Disease Control and Prevention explains that nontherapeutic, low-dose antibiotic use in farmed animals is contributing to antibiotic resistance and should be kept to a minimum. (They actually recommend that antibiotic use to promote growth should be completely phased out.)[2]

In response to these concerns about biosecurity, I said, when talking with the farmers: "I would be happy to buy my own hazmat suit and whatever other protective gear I might need so that I won't contaminate the farm."

One farmer replied, "That's impossible. That won't work."

"Oh! Can I ask why not?" I asked.

"It just won't; we have very strict protocols for safety and we can only have people coming on the farm who need to be here. I'm sorry, I'm going to have to go now. Best of luck."

During one phone conversation, I asked the farmer, "Can you give me a sense of what biosecurity means?"

Flustered, he replied, "It's, you know, the safety and cleanliness of the farm. It's important for national security and the security of the food supply. You know what, I've gotta get back to work." And he hung up. Throughout many of these conversations, I found that *biosecurity* is a complicated and ambiguous buzzword used in supporting the denial of access to spaces of agricultural production under the real or perceived guise of food safety and disease prevention.

Certainly, food safety and the prevention of disease are not issues to be taken lightly; especially in a highly industrialized food system, food safety, cross-contamination, and the rapid spread of disease among farmed animals kept in close contact are threats to human health, environmental safety, and animal well-being. However, the discourses of biosecurity operating in spaces of animal agriculture do not tend to reflect the casualness with which workers in the auctions and the farm I visited (and in written accounts of dairy farming that I reviewed) moved to and from the farm. I was hard-pressed to find documentation in dairy-farming literature or in the spaces I visited during my research of the boundaries of these spaces being protected from threats to biosecurity—beyond routine practices such as pasteurization of the milk, prophylactic usage of antibiotics, and the use of iodine to clean cows' teats in the milking process.

AG-GAG LAWS

I was also surprised to find, through my preliminary research, that a collection of laws at the state level—termed *ag-gag laws*—criminalize the documentation of the activities of agricultural facilities (particularly those raising animals for food). Each ag-gag law has slightly different language and coverage, but they generally prohibit video, audio, and photographic documentation of activities in spaces of animal agriculture. At their heart, they are anti-whistleblower laws that make it

difficult even for employees who might report gross animal welfare violations to document and file a complaint. But the more insidious intention behind these laws is the prevention of animal rights activists, journalists, and other concerned citizens from documenting the conditions
under which animals are raised for food. In fact, while I was in the midst
of writing this book, the first prosecution under an ag-gag law was filed.

On February 8, 2013, a living, immobile cow, termed a *downer* by the
industry, was moved with a tractor across the grounds of the Dale T.
Smith & Sons Meatpacking Company in Draper, Utah.[3] Amy Meyer
stood on the side of the road next to the slaughterhouse and watched.
As a consumer interested in food production practices, Meyer had
heard that the animals moving into the slaughterhouse could easily
be seen from the road at this particular packing plant and went to see
for herself. Meyer, who stood on the public easement next to the road,
was a hundred feet from the slaughterhouse with a fence and field in
between her and the structure. From there, using her phone's video
camera, she recorded what she saw.[4]

Similar video footage showing collapsed but conscious animals
moved with forklifts had been used by the Humane Society of the
United States to prosecute a California meatpacking plant in 2008 in
the largest US meat recall in history, where it was discovered that sick
animals were entering the food supply and, specifically, the national
school lunch program.[5] As Meyer stood there filming, Bret Smith, the
manager at the meatpacking company, approached her in a pickup
truck and informed her that she was not permitted to film the slaughterhouse. Capturing the exchange on her own video camera, she replied that she was under the impression that where she was standing
was a public easement and that it was legal for her to film the slaughterhouse from there.[6] Smith replied that she was filming the plant from
private property and that he would call the police. Meyer responded
that she would be willing to move from the roadside if the police told
her it was not a public place.

SMITH: You cannot videotape my property from public property, so
I'm asking you to stop. Now I'll leave and call the cops . . . if

you have something to ask me about my business, why don't you have the balls to come and ask me? We're running a legitimate business over there and you guys have no business recording me from anywhere.

MEYER: Why are you concerned with being filmed if you have no problem, if you think this is a legitimate business?

SMITH: If you, uh, read the rights here, the laws in Utah, you can't film an agricultural property without my consent.

MEYER: That's correct, on your property.

Smith called the police. When the police arrived, there was some confusion about the actual content of the laws. The police officers began by saying that she was not being arrested or charged with anything, that they just wanted to ask Meyer some questions. Meyer asked the officer to verify that where she was standing was a public easement. He verified that it was. Meyer asked repeatedly if she was being detained and the officer replied that she was not. Meyer stated, "I don't need to answer any of your questions. I have done nothing illegal. I am not suspected of anything illegal. I'm on a public easement next to the street. . . . I'm not going to talk to you. I'm done talking with you. When my lawyer calls, I'll let you talk to him."

The police officer eventually stated that Meyer was free to go and informed her that prosecutors would determine whether she should be taken to court after they "screened charges."

Meyer was charged with agricultural operation interference under Utah's HB 187. This was the first prosecution under a state ag-gag law. Due to overwhelming public outcry, the charges were dropped twenty-four hours later.

This event made me realize how easily I could have been prosecuted under an ag-gag law if I had used deception to obtain employment in the dairy industry in a state with an ag-gag law on the books. Washington State has not passed an ag-gag law, as of the writing of this book; however, one was proposed there in January 2015 and was met with serious opposition. Whether or not Washington passed an ag-gag law, it became clear through my research that the broader climate of these laws passing across the United States creates a culture of fear for institutions like the university (and individual researchers) that could possibly be the target of litigation under future bills.

The first ag-gag law was passed in Kansas in 1990 and prohibited photography or filming, property destruction, and/or the theft of animals from agricultural or university medical research facilities. Other midwestern states passed similar laws through the 1990s and into the decade following. Some differed in scope and coverage. Some of the things that are covered include stipulations that only law enforcement agents are permitted to investigate animal cruelty (in the case of Arkansas's 2012 law).[7] Or Iowa, for example, where it is now a crime to enter an agricultural production facility under false pretenses.[8] The implications of these laws span beyond just the control of outside researchers or investigators. Missouri approved a bill stating that employees filming an act of cruelty must report it to the authorities within twenty-four hours—a law that, on the surface seems to be concerned with stopping certain actions immediately, but in effect makes it illegal to build a more substantive case of long-term cruelty charges against a facility.[9] Utah's 2012 bill, prohibiting the recording of agricultural activity, was used just a year after it passed to bring charges against Meyer for taking video of a slaughterhouse from a public road.[10]

Future iterations of ag-gag bills may broaden their scope, as recently proposed bills indicate.[11] North Carolina's Commerce Protection Act, for instance, is not specific only to agricultural spaces or medical research facilities but could apply to limiting the distribution of information in many industries.[12] Bills such as these could be used, for example, to prevent opposition to the environmentally deleterious effects of fracking and other extractive industries.[13] Journalist and activist Will Potter has done extensive work to bring the effects of laws like ag-gag to light. He argues, in his book *Green Is the New Red*, that laws like these are meant to silence dissent and the distribution of information about consumer-related industries and practices, and they reflect a stark prioritization of corporate interests. Those proposing the laws at the state level have close ties to agricultural industry interests, and many of the bills are endorsed by industry interest groups (e.g., California's bill, which was defeated in 2013, was sponsored by the California Cattlemen's Association).

What I didn't know at the beginning of this project was that ag-gag laws are part of a much broader, more insidious sweep of law making at the federal level. Though the first iterations of ag-gag bills took shape in the 1990s, the political climate after September 11, 2001, made it possible for bills restricting access to places of animal use (e.g., agriculture and research laboratories) to become more ubiquitous across the country.

After September 11, 2001, the Uniting and Strengthening America by Providing Appropriate Tools Required to Intercept and Obstruct Terrorism Act—better known by its abbreviation, the USA PATRIOT Act or, simply, PATRIOT Act—was passed to remove restrictions on intelligence collection by US law enforcement agencies. *Terrorism* is defined by the US government as "the unlawful use of force and violence against persons or property to intimidate or coerce a government, the civilian population, or any segment thereof, in furtherance of political or social objectives."[14] Among other changes to federal law, the PATRIOT Act made it easier for law enforcement agencies to detain immigrants suspected of engaging in "terrorist" activity and, importantly for the subject of this book, it expanded the definition of *terrorism* to include the category of domestic terrorism. The PATRIOT Act defined *domestic terrorism* as activities that "involve acts dangerous to human life that are a violation of the criminal laws of the United States or of any State" or activities that "appear to be intended to intimidate or coerce a civilian population; to influence the policy of a government by intimidation or coercion; or to affect the conduct of a government by mass destruction, assassination or kidnapping" and if the activity occurs "primarily within the territorial jurisdiction of the United States."[15]

The post-9/11 passage of the PATRIOT Act paved the way for other legislation that represses and restricts rights to privacy and free speech. In 2006, the federal Animal Enterprise Terrorism Act (AETA) was passed as an amendment to the Animal Enterprise Protection Act

(AEPA) of 1992. "The AETA expanded the AEPA to include both successful and attempted conspiracies. It also prohibits intentionally placing a person in 'reasonable fear' of death or serious bodily injury while damaging or interfering in the operations of an animal enterprise."[16] Drawing on the PATRIOT Act's category of domestic terrorism, the AETA targeted those involved in direct action tactics (economic damage, threats of damage, etc.) against animal enterprises (farms, laboratories, etc.).[17]

Will Potter argues that, instead of responding to a real threat against human safety, efforts to condemn animal rights and environmental direct action as terrorism serve corporate economic interests and perpetuates what he terms a "Green Scare." Potter's Green Scare harkens back to the communist Red Scare and is characterized by intimidation tactics involving the use of terms like *ecoterrorist* and *animal rights extremist* to promote fear and silence dissent in all its forms, as well as advance a political-economic agenda that privileges businesses that profit from the exploitation of animals and the environment. These efforts to criminalize dissent may come in the form of silencing illegal activity, but this culture of fear around dissent may also silence legal resistance in the form of nonviolent civil disobedience and other forms of activism, film, and journalism, as well as potentially deterring academics from speaking out in alliance with the political perspectives of the animal liberation or environmentalist movements. And, importantly for this project, they create the legal climate in which state laws, like ag-gag, can be passed and upheld. Once I made myself familiar with this legal climate, I understood much more fully the context in which the university IRB would be concerned with liability issues related to obtaining employment on a farm under false pretenses.

But I also understood the sobering effects of laws like ag-gag and the AETA in terms of making it harder to uncover and oppose violations of animal welfare and distribute information that would inform the public about what occurs behind increasingly closed doors. Film and photographs, taken by undercover investigators, have been used to marshal criminal cruelty charges against individuals and corporations, as well as to educate the public about the acts of violence to which animals and human workers are subjected in food production. Often, these under-cover efforts are the only access the public has to information about meat, dairy, and egg production practices. Internal whistleblowers for the industry are particularly important, as laws protecting animals in food production are notoriously weak and insufficiently enforced.

Interestingly, in the years since I conducted my research, there has been an emerging trend in highly controlled industry transparency that suggests a new kind of engagement with what is visible, how, and to whom. Jan Dutkiewicz's work and Timothy Pachirat's most recent project involve an analysis of Fair Oaks Farms in Fair Oaks, Indiana.[18] Fair Oaks is an industrial-scale dairy and pig farm that charges admission for tourists to come and observe the process of raising cows for dairy and pigs for meat. In what Pachirat describes as "the commodification of transparency," Fair Oaks offers a carefully curated experience to paying visitors—one that makes visible certain aspects of the production process (in effect, normalizing certain kinds of violence) while continuing to conceal others. Paired with the political climate in which ag-gag laws emerge, this kind of curated view of animal agriculture will be something to watch.

ANIMAL WELFARE LAW IN THE UNITED STATES

To understand the broader political context for the lives of animals in the dairy industry, it is necessary to look into the animal welfare laws governing how animals live and die for the production of milk. I found that looking back just over the last hundred years of US history, there is a cyclical process wherein whistleblowers, journalists, or scholars expose welfare violations, the public is horrified and demands action, and the government is pressured to pass more protective legislation. This cycle comes in waves in which interest among investigators and the public ebbs and flows—but the pressures exerted by industry power and wealth are continuous.

The Animal Welfare Act, a federal law enforced by the USDA designed to "protect certain animals from inhumane treatment and neglect," does not include animals used for "food, fiber, or other agricultural purposes." The Humane Methods of Slaughter Act (HMSA), initially passed in 1958, is surprisingly the *only* federal law that includes farmed animals in its coverage. Oddly, certain species killed for food are not even included in the HMSA, such as birds, fish, and rabbits, which is especially concerning, considering that more birds and fish are killed each year than all other farmed animal species combined. The HMSA was drafted and passed after mounting discomfort with the meat industry among consumers and a growing understanding that the meat industry operated inhumanely in a number of ways (both for animals and humans). These concerns were built on the knowledge of dangerous conditions for workers in the meat industry and unsanitary conditions for the production of food, which had been brought to the public's eye much earlier, in 1906, with Upton Sinclair's groundbreaking work of fiction, *The Jungle*. Then, in the 1950s, news reports spread about what was framed as cruelty to animals in slaughterhouses in the United States. The realization that animals were being treated so badly in the food industry, paired with the repeated exposure to workers' experiences in the meat industry and unsanitary conditions for food production, fostered a political climate in which the HMSA was passed.

The HMSA was designed to improve animal welfare, working conditions, and food safety in the meat industry. Interestingly, though, there is particular emphasis in the text of the HMSA on the economic benefit associated with the improvement in the quality of the meat. Animals not injured or mistreated during the process of slaughter provide more sellable meat; injured animals produce flesh that is bruised and has to be trimmed more extensively, resulting in more waste and loss of profit.

The HMSA of 1958 "required that livestock be rendered insensible to pain by a blow, gunshot, or electrical or chemical means that is rapid and effective before shackling, hoisting, casting, or cutting." Compliance with the 1958 law was *voluntary* for all producers not selling their products to the federal government. This meant that meat sold to the general public was not required to be produced under these standards. The Humane Methods of Slaughter Act of 1978 (an amendment of the 1958 law) is the current law governing the slaughter of some animals for food. In 1975, Peter Singer published his now-classic *Animal Liberation*, which shocked the public with its unflinching exposé of the conditions under which animals were raised for food. This generated a flurry of concern about the welfare of animals in the food industry.

As a response to this new wave of concern about animal welfare, in 1978, the HMSA was amended and made *mandatory* for all USDA-inspected slaughter facilities. The law outlined two approved methods of slaughtering animals. The first was that animals must be "rendered insensible to pain" prior to slaughter by a gunshot, blow, electric shock, or efficient chemical exposure. The second method of slaughter deemed acceptable was "ritual slaughter," in which animals have their carotid artery cut and are rendered unconscious by a rapid loss of blood. Common examples of animals killed by ritual slaughter are those slaughtered for Kosher or Halal purposes. The 1978 amendment also added a note that the humane *handling* of animals should be considered during the process of slaughter. The current HMSA applies to adult bovine animals, calves, horses, mules, sheep, and pigs but excludes birds, rabbits, and fish.

In 2001, *Fast Food Nation* by Eric Schlosser was published and quickly became a *New York Times* bestseller. Schlosser recalled many of the same and similar details of slaughterhouse workings as did Sinclair in *The Jungle*. This time, though, the focus was on the historical development of the fast food industry and the unsanitary, unsafe, and unethical circumstances under which food was produced for this industry. Also in 2001, an article titled "They Die Piece by Piece" by Joby Warrick was published in the *Washington Post*, which documented humane-standards violations, exposing inhumane conditions at a vast number of cow slaughter facilities. Cows were witnessed being skinned alive and having their legs and tails cut off and their bellies cut open while still conscious. The article revealed that these violations were not unusual but, in fact, commonplace.

The public outrage raised by Warrick's article and by *Fast Food Nation* pushed the government to respond to accusations of lax enforcement standards and general negligence. The George W. Bush administration responded in the 2002 Farm Bill, which suggested greater enforcement of the HMSA standards. This encouragement of enforcement was meant to ease public alarm over conditions in the meat industry as well as make the meat industry operate more efficiently.

This brief historical account of federal farmed animal welfare legislation over the last century uncovers the ways in which journalists, investigators, and academics have been integral in raising public awareness about how animals are treated in food production. This public awareness, in turn, generates pressure on lawmakers to draft and/ or further enforce farmed animal welfare legislation. These changes are then made into law, the original concerns about welfare and food safety are then forgotten or assuaged, and things generally continue on as they were before the public outcry. Indeed, despite decades of this cycle, federal protections remain weak or nonexistent, laws are insufficiently enforced, and there are still no federal welfare laws governing the lives of farmed animals.

Even more concerning than the lack of federal legislation is the way state laws operate. State anticruelty and animal welfare laws are designed to protect certain animals from what gets characterized as "cruel" treatment. However, the details of these laws are broad and apply to all types of animals, which cause a distinct lack in specifics relating to farmed animal welfare.[19] While they generally include farmed animals, there is no enforcement system to ensure that these laws are followed. Additionally, attorneys David Wolfson and Mariann Sullivan point out that "the burden of proof on the prosecution is very high, that is, beyond a reasonable doubt."[20] Abusers of farmed animals are generally not prosecuted for violations because the "burden of proof" requires proof of *intention*. Wolfson and Sullivan outline an example in New Jersey, which illustrates the difficulty in prosecution: an egg company was charged with violating anti-cruelty laws for throwing sick, but living, laying hens into a garbage can filled with dead hens, and leaving them there to die. Because the prosecutor could not prove that the employee had "knowingly" discarded living hens, the case was dropped. In addition to the difficulty of prosecuting producers for cruelty to farmed animals, those convicted are usually required to pay only limited fines. Wolfson and Sullivan say, "for example, Maine has a maximum fine of $2,500, Alabama and Delaware have a maximum fine of $1,000, and Rhode Island has a maximum fine of $500, for general cruelty to animals."[21] These fines are such trivial amounts that it is often easier for corporations to just pay them rather than make any systemic change in their company practices.

Another major obstacle to the enforcement of anti-cruelty laws and the HMSA enforcement is what David Wolfson terms Customary (or Common) Farming Exemptions (CFEs). These exemptions are laws passed at the state level that grant the animal agriculture industry exemptions from animal cruelty laws. Erik Marcus, in his book *Meat Market*, writes, "The majority of states have put CFE laws on their books ... Using words like 'common,' 'customary,' 'accepted,' and 'established,' CFE laws allow any method of raising farmed animals to continue, no matter how cruel, so long as it is commonly practiced within the industry."[22]

What these laws mean in practice is that as long as a particular form of inhumane treatment of animals is deemed "common" by the industry itself, a company cannot be prosecuted. Wolfson and Sullivan conclude that, "as a result, in most of the United States, prosecutors, judges, and juries no longer have the power to determine whether farmed animals are treated in an acceptable manner. The industry alone defines the criminality of its own conduct."[23] Acceptable practices include severing the tails of pigs and cows, castrating young animals, and cutting chickens' beaks off with a hot blade—all done without anesthetic. During the slaughter process, acceptable practices include using electrified prods to force animals to move, as well as allowing a certain number of animals to be conscious through the slaughter process. These are all examples of routine practices that occur regularly in the industry and are exempt from anticruelty laws, but even if these particular practices were recharacterized as "cruelty" and abolished, all of the other dimensions of breeding, raising, and slaughtering animals for food that are outlined in this book would still remain acceptable as the status quo.

To give an example of the kind of language used in these laws, Washington's CFE law reads, "Nothing in this chapter applies to accepted husbandry practices used in the commercial raising or slaughtering of livestock or poultry, or products thereof or to the use of animals in the normal and usual course of rodeo events or to the customary use or exhibiting of animals in normal and usual events at fairs."[24] Farming exemptions like this one operate in many states to forgo protection of farmed animals in their daily lives. As long as enough producers agree on or conduct similar practices, these practices become routinized and normalized and power is granted to the industry itself to define how animals are and should be treated.

Anticruelty laws, though, are problematic not merely because they exclude or exempt certain animals or practices; rather, the very concept of cruelty in conversations about human-animal relations should be scrutinized. According to animal studies scholars Will Kymlicka and Sue Donaldson, the concept of cruelty operates to normalize common practices by the majority group and to villainize those practices that fall outside of what the majority group counts as normal or acceptable. They write that "majority practices are inherently immunized from moral and political scrutiny. . . . Indeed, this is a central purpose of so-called animal welfare laws. Their goal is not to protect animals, but to provide legal cover to those who benefit from harming animals."[25]

For Kymlicka and Donaldson, and for political scientist Claire Jean Kim, this point is illustrated in a variety of culturally and racially charged examples of animal use.[26] Take, for example, the growing public outcry in Western countries like the United States and the United Kingdom over the cultural practices of raising dogs for meat in parts of China, Korea, and the Philippines. The Yulin Dog Meat Festival in Yulin, Guangxi, China, especially, has come under fire for what gets framed as a "barbaric" practice that must be stopped.[27] Each year, in the months leading up to the festival, my social media feeds are, with increasing intensity as the date nears, flooded with articles, petitions to stop the festival, and declarations of disgust over what people inevitably call the "cruel," and "barbaric" practice of eating dogs. These outpourings come both from vegan and vegetarian animal advocates *and* from broad swaths of the general public who regularly engage in the act of eating species like pigs, cows, chickens, and turkeys, for example. Kymlicka and Donaldson point out that, "from an AR [animal rights] perspective, eating dogs is no better or worse than eating pigs: they both violate the fundamental rights of animals to life and liberty. The broader public, however, endorses the principle that humans do have the right to harm and kill animals for our benefit, so long as we avoid 'cruel' or 'unnecessary' harm. It is this principle that opens the door to bias since perceptions of what is cruel or unnecessary are culturally variable. . . . The idea that it is cruel to eat dogs and horses but not cruel to eat pigs and cows is a cultural idiosyncrasy . . . [and] the public mobilizes around these distinctions, often to the detriment of minorities."[28] The use of terms like *barbaric* and *cruel* here operate not only to racially and culturally exclude certain groups but also to cement Western majority practices (farming cows, pigs, and chickens) as *civilized, normal,* and *acceptable.* Anticruelty and animal welfare laws in the United States (as well as other places) help to entrench these divisions and normalize widespread majority practices.

These nonexistent, lax, and insufficiently enforced laws for farmed animals, paired with increasingly repressive federal and state legislation preventing the collection and distribution of information about spaces of agricultural production, create a complex and difficult legal climate in which to study and write about animals used for food. For those concerned with animal welfare, food safety, and transparency in food production, it is especially important not only to understand these laws and the political agendas that make them but also to be informed about their impacts on the real, living beings whose lives and deaths they are designed to conceal. The nature of how anticruelty and animal welfare laws operate also prompts questions about whether law has the potential to make meaningful change in how humans relate to and treat other species; legal scholar Maneesha Deckha, for instance, worries over the entrenched anthropocentrism in law and the way legal systems often further entrench species hierarchies and human supremacy.[29]

Within this legal context—and the industry's culture of barring access to sites of food production—I was especially grateful when Homer Weston at Ansel Farm in western Washington allowed me to come and visit his five hundred–cow dairy.

FIGURE 3.1 The heifer with ear tag #6490, Ansel Farm (Western WA)

3

THE SMELL OF MONEY

My dad was visiting Seattle when I finally found a farm—Ansel Farm in western Washington—that would allow me to visit. On the phone, Homer Weston was cheerful and receptive and suggested I come to visit his farm the next day.[1] I was, at this point, shocked. My days had become punctuated by uncomfortable phone conversations with farmers in which I explained that I was a university researcher trying to better understand the dairy production process, and I was repeatedly denied. I had become accustomed to making these calls and accustomed to being turned away. It had, honestly, become a daily routine, done over a morning cup of coffee, or an afternoon cup of tea. I'd sit down at my desk, get my notebook out, start by calling one farm, jot down the responses I got, get turned away, and move on to the next farm on my list. Every few days, I would do more research on farms in California, Washington, and even Oregon, making new lists of places to call, all the while keeping meticulous track of the correspondences so as to follow up at the appropriate time or so as not to accidentally call a farm again that had given me a firm "no."

"Oh!" I said, unable to conceal my surprise over the phone when Homer agreed to a visit. "Really? I can come tomorrow?"

"Sure, can you come in the morning, in between milkings? We have about five hundred cows here at the farm—most of them are in the milking string, so we have to stick to our milking schedule to keep things running on time. We make our own cheese right here on the farm. Lots of varieties. You can see when you get here."

"Great! Thank you. You name the time and I'll be there!"

"How about 10 A.M.?" Homer proposed.

"10 A.M. it is. Oh! Can I bring my dad?" I asked.

"Of course. Just park by the shop and I'll meet you there. See you tomorrow." Homer replied, and then hung up the phone.

I was nervous and excited, and tossed and turned that night, thinking about how I would finally be able to begin my fieldwork. I felt relieved, too, that my dad was in town and could come along. He's a laid-back guy—easy to talk to, with a good sense of humor and an unpretentious way about him that puts people at ease.

The next morning, we headed out to the farm, which was a couple of hours outside of Seattle. We had agreed to visit midmorning, when there was a lull between milkings so that Homer would have time to show us around. Because of this schedule, we realized we would have the afternoon free and I did some research before we set out to see what was in the general vicinity of the farm. There was a farmed animal auction yard nearby, and it just so happened that they were having a dairy market sale that afternoon, so we decided we would attend the auction after our visit to the farm.

Almost exactly two hours after leaving home, we pulled into the gravel driveway at Ansel Farm and parked in front of a small building—the cheese shop. When Homer told me over the phone that there were about five hundred cows in Ansel Farm's milking herd, I expected to arrive at a large farm. But Ansel Farm looked small and modest set against the rural landscape. The farm buildings were laid out in a small area with the cheese shop just in front of the large barn and pens where the animals were housed, which was adjacent to the cheese-making buildings and the storage areas. Behind the barn were enormous piles of silage (a feed ration composed of fermented grains and grasses) covered with plastic sheeting weighted down by truck tires to aid in the fermentation process. Out beyond the barn stretched the farm's acreage: fields of feed crops, and beyond those fields, a manure lagoon—a large pit where farmworkers would dump liquefied manure from the farm.

An elderly man with a long white beard and overalls sat on a porch swing outside the cheese shop. When we got out of the car, he approached—he had been awaiting our arrival.

"Are you Katie?" he asked with a half-smile.

"Yes! Homer?" I walked up to him and shook his hand as he nodded. "Thank you so much for taking the time out of your day to show us around. Shall we fill out the paperwork I mentioned first?" As an IRB requirement, I provided Homer with a consent form to use our interview and tour in my research with his (and the farm's) identity anonymized.

"I hope I'm not signing my life away here!" he joked, as he read and completed the form.

After the paperwork was complete, Homer stood up, and beckoned for us to follow him. "Come on, I'll show you the cows first."

As we walked up the gravel drive, I asked Homer about his background: "Can you tell us a little about your experience with dairy farming? Have you been doing it long?"

"Oh sure. Yeah, I'm what you call a lifer," He chuckled. "I grew up on a dairy farm in Wisconsin. It was a family business and it was my grandfather's before it was my father's. We did everything there on the farm—grew crops for feed, reproduction, milking, you name it. But it's hard to keep a small farm going these days and so now I work here for the folks that own this place."

"How big is this farm here?" I asked.

"Well, as I said on the phone, we have about five hundred head and this here's about a hundred sixty acres. And then we rent another hundred fifty acres where we move the dry cows and raise the heifer calves."

We came to a large barn with a series of divided areas of pens containing herds of cows.

"This is the maternity pen." Homer pointed to the enclosure closest to us. The maternity pen was filled with large black-and-white Holsteins and small creamy-brown Jerseys. All of them had tags punched through their ears and the farm's brands on their hips. The pen was barren with a dirt, manure-covered ground and troughs of feed and water.

As we approached, a small Jersey heifer with ear tag #6490 walked up to me and stretched her head out, neck extended over the fence. Startled, my instinct was to step back, but Homer said, "Oh, you're fine. She's just curious." Homer had a soft spot for Jersey cows. As we stood there looking at the maternity pen, he said, "There are good qualities about both breeds—I mean, the Holsteins and Jerseys—but there's nothing sweeter than a little Jersey heifer. Look at those eyes."

Gently, she reached out her long gray and pink tongue and licked my arm. Her large brown eyes, long eyelashes, and fuzzy cap of reddish brown hair were hard to resist, and I reached out to scratch her neck and behind her ears. She licked me again. Curious, other cows started to crowd around and, soon, my dad and I were petting and scratching and being licked by a small herd of cows who crowded at the fence to reach us. Homer chuckled and patted the cow nearest him. My hand brushed the ear tag on the small Jersey heifer. Homer said, "So do you know about the ear tag numbering system? I can explain it."

"No, I don't. I'd be really interested to hear." I looked at the heifer with ear tag #6490. She had a large yellow tag on each ear, with "6490" written in large numbers—no other markings. I glanced at the cow next to her and her ear tags had three rows of numbers—the top and bottom numbers were in smaller print, and the middle was in large print.

"So, if you look at 6490 there, that's the simplest numbering system. That's just her ID number. But then there's other numbering systems with more detail. Look at that one there," he pointed to the Jersey cow standing next to her. "The top number is the ID number for the cow that birthed her. The bottom number is the bull—the sire. And the middle, big number is the ID for this particular cow. This way we can easily keep track of which bulls and cows produce the best cows for milking. Some ear tags only have the one number, like #6490, some have all three numbers, some have the sire's name instead of a number (lots of times sires have names), and different configurations like that."

"Why are they different? I mean, why aren't they all standardized here?" I asked.

"Well, it just depends when you get them. Ear tagging is done when they're real young, so we often buy heifers who've already been tagged, if they weren't born here on the farm. So, you get a mix of styles. But you at least have the most important info—the number of that animal—on all of them."

In later research, I read more about ear tags: they are generally plastic or metal tags with identifying numbers and sometimes barcodes that, as Homer had explained, can contain an animal identification number, the herd, the sire and dam from which the animal was bred, the farm, and/or any identifying characteristics the farmer wishes to include. Animals may have ear tags on one or both ears. Electronic ear tags are gaining popularity and are scannable by machine in more automated systems where computers may track the activity of each animal. Ear tags are generally attached to the animal's ear with an applicator that punches the tag through the ear. The tag punches a permanent hole in the animal's ear, and this trace of identification and mark of ownership can be read on the body long after the ear tag is gone. (Sadie, for instance, had a hole permanently remaining in each ear years after she left the dairy farm.)

"Interesting," I said. "That reminds me—I've read about tail docking as a common practice, but it doesn't look like any of these cows I can see have had their tails docked. Do you do any tail docking here?"

"You're right. We don't have any cows here who've been docked. That's not something we do. It really varies farm to farm on whether folks'll decide to dock the tails. The reason a lot of people do it is that some people think it makes the whole process cleaner," he trailed off for a moment and then continued, "Sometimes the tail gets manure on it and that can sometimes get on the udders and so some people think that if you remove the tails you remove that contamination risk and also the risk of causing mastitis. But we find here that it's just fine if you clean off the teats properly before and after milking. Plus, cows need their tails to swat away flies."

In later research, I explored tail docking in more depth. Docking is performed primarily on females either at weaning or just before or after a cow's first calving. Methods for tail docking include using an elastic band, pruning shears, a cauterizing docking iron, or an instru-

ment called a burdizzo, which is a pliers-like device used to break the
tail bones.[2] The elastic band is the most common method of tail docking in the United States and involves an elastic band being applied
tightly around the tail with the effect of making the majority of the
tail necrotic. Three to seven weeks after banding, the necrotic portion
of the tail will fall off, or it might be cut with pruning shears by the
farmer.[3] A cautery tail docker can be used to cut the tail and cauterize
the stump. And a burdizzo can be used to break the tail, after which
shears would be used to cut the tail off below the break.

Tail docking causes the animal both short- and long-term, acute
and chronic, pain—the tail is composed of bone that must be broken
in the docking process—and cows with docked tails suffer the same
"phantom limb" experience as humans who have had a limb ampu-
tated.[4] There is no evidence that tail docking improves health, reduces
instances of mastitis, or improves milk quality.[5] The American Veterinary Medical Association, in fact, "opposes routine tail docking of
cattle. Current scientific literature indicates that routine tail docking
provides no benefit to the animal, and that tail docking can lead to
distress during fly seasons. When medically necessary, amputation of
tails must be performed by a licensed veterinarian," and docking is now
illegal in a number of US states.[6] Of course, cows' tails serve important
functions for their comfort, health, and social relationships. Cows use
their tails to communicate with others and signal estrus activity and
to protect themselves against flies, and their tails are also important
for temperature control, as they help to keep them cool in summer
months.

Homer gestured at the maternity pen and continued: "So, there are
about sixty in the maternity pen here right now who'll be birthing in
the next couple of weeks. That one you're petting there [with ear tag
#6490], this is her first time. She's just about two years old."

From research I had already done on dairy production, I knew that
it varied from farm to farm how many times cows would be impregnated before they were deemed "spent" and sent to slaughter. Some
farms keep their cows for three to five years; others longer. I asked

Homer about Ansel Farm's practices: "How many cycles of pregnancy and milking will she likely go through before she's worn-out?"

"Oh, it really depends. They get pregnant at about fifteen or sixteen months, and then are birthing for the first time at around two years old. I'd say on average we keep them about five or six years—so three or four calvings . . . sometimes more than that if she's doing well."

"Can you tell me what the typical year in the life of a dairy cow at Ansel looks like?" I jotted down notes in my notebook and then looked back up at Homer.

"OK, sure. Well, let's take this one here," he said, pointing at the heifer with ear tag #6490. "Since she's just starting out. She'll deliver a calf here in the next couple of weeks. And then she'll move into the milking string, which I'll show you next, and we'll milk her for about three months before she's ready to be artificially inseminated again. Then we can actually milk her for most of her pregnancy— the gestation is nine months (like humans), so we usually stop milking them about sixty days before they're ready to deliver. That's called 'drying off.'"

My dad interjected, "So they can be milked for about ten months out of every year?"

"That's right. Sometimes nine, but usually we can do ten—cows will produce milk for three hundred days of the year." Homer continued: "So, after drying off, we take them to another farm, where they eat and rest and get ready to deliver. We move 'em back here into the maternity pen when they're a couple weeks away. Then they deliver and then they move on into the milking string again. It's a cycle where there's a calf a year."

"What happens when the pregnant cows go into labor? Can you describe the process?"

"Sure. Well, we move them into the maternity pen when they're within a few weeks of birthing. When they go into labor, we isolate them and keep an eye on them. Occasionally, if it happens at night we might miss the event entirely and we come in the morning and there's a calf, but mostly we know when they're nearing delivery time and we're here to make sure things go smoothly. Hopefully the birth is an easy one and we don't have to intervene too much—we like it best when they can, you know, do their own thing."

"What happens to the calves when they're born?"

"Sure, yeah. We take the calves out pretty quickly—you know, within a day usually—and we have a separate farm where we raise the heifer calves until they're ready to come back here, or they get sold to another dairy farm. It depends on our numbers and needs at the time, really."

"Why do you remove the calves so soon after birth?"

"It's better that way. We need to separate them for the good of the cow and calf. The longer they bond, the harder the separation is. You know, it's kind of sad. Even when we remove the calves so quickly, the cows'll bellow for the calves—like they're looking for them—for a couple of weeks a lot of the time. So yeah, it's just better to get it over with quickly so they don't get too bonded."[7] In my later research, I found that this trend of early removal of calves is commonplace in the industry. The USDA Animal and Plant Health Inspection Service documents that roughly one-fourth of dairy farms in the United States remove calves within one hour of birth and more than half (57.5 percent) separate calves and cows between one and fourteen hours after birth.[8]

"What about the male calves? What happens to them?" I asked.

"Since we can't use them on the farm, a cattle buyer comes to the farm and buys them."

"And is that to raise them for meat? Or breeding?"

"Meat, mostly," Homer replied.

"Are they raised for veal? I read an article about local veal producers partnering with small dairy farms to provide a local source of veal in the Pacific Northwest." I pressed on.

Homer cleared his throat and shifted on his feet. "We have nothing to do with veal here and we don't want to." I sensed that I should not push further. As I talk about later in the book, there is a direct, yet uncomfortable, link between the production of dairy and veal. Many male calves born into the dairy industry, who are of no use to dairy

production, are routed into veal. Yet many dairy farms do not publicize
this link, given decades-long public concern over the ethics of veal.

I changed the subject: "OK, so the cycle of reproduction and
milking—when does it end? How do you decide that they're ready
for slaughter?"

"That's determined by a number of things, and it differs from farm to
farm, actually, depending on how long they want to keep the cows—"

I interrupted: "Oh yeah, I read that larger, industrial-scale farms will
send cows to slaughter more quickly—sometimes after just two or
three cycles. Is that right?"

"Yes," he paused, "Usually. It really depends. Larger farms do tend
to move them through more quickly. When I was growing up in Wisconsin on my family's dairy farm, we kept the cows longer—here, we
keep them till they're six or seven. And then it depends cow to cow,
too. You know, you get some cows that handle milking better and then
others that don't so much or they get diseases. So you have to make
decisions based on each animal."

"What kinds of diseases would make a cow's productivity decline?"
I asked.

"Well, there's a few common things. First, there's just a general
decline—that's not related to disease—so, they just don't produce as
much milk as they age, or they don't get pregnant as easily after a while.
And then there's infertility—you know, both heifers or cows who just
don't get pregnant at all or have trouble. Or they used to, but after
a few rounds, it's hard to get them pregnant. But then there's things
like mastitis that we deal with. Mastitis is probably the most common.
It's a disease, an infection of the udders. They get inflamed—red and
swollen—and causes the cow discomfort. If a cow gets a bad case of
mastitis, it can be hard and expensive to treat. So sometimes that's a
reason [to send her to slaughter]."

I interjected: "Oh yeah, I read about mastitis. I think it also comes
with sort of flu-like symptoms, right? Body pain, fever, fatigue. Do you
see that in the cows?"

Homer nodded. "Uh huh. The cows definitely don't feel good when
they get it. And milking is hard because the udders are pretty tender

if an infection is coming on. Although, we try to catch it early so we aren't milking cows with mastitis. It can cause bleeding and sometimes there's pus from the infection. We don't want to get infected milk in the products we're selling."

My dad, who had wandered ahead to look at another pen of cows, returned and asked, "What else would cause a cow to be sent to slaughter?"

"Lameness is also common. That's when cows develop a bad limp or can't put weight on a leg. You see lameness most when cows are housed on concrete." Homer pointed into the barn, "See, there, the barn is cement flooring. We have cement in there because it's so much better for keeping things clean. It's easy to wash and clear out the manure. But cement is hard on them. And sometimes it can be slippery—if you can kind of see from here," he said, pointing again at the cement floors, "there's grooves in the cement. That's to keep the cows from slipping. It gives more traction. If you have floors without grooves, those get extra slippery and the animals can fall and injure themselves, which is one thing that can make them lame. Injury, I mean."

I nodded, urging him to continue.

"But then it's also just hard on their legs and hooves to stand on cement, so we try to make sure they have options." He pointed to the outdoor pen we were standing in front of. "This dirt is much softer and so they can get a break from the cement and some fresh air. That's important for keeping a healthy herd." Homer stroked his beard and paused.

"So, do they spend most of their time outside year-round? I know how muddy the Northwest can get in winter with all the rain." I pointed at the dry dirt and manure ground of the outdoor pens.

"Yeah, it gets muddy in the winter and they'll spend more time inside, out of the rain, but they do still come out into the pens. But, yeah, we do worry about lameness when they spend more time in the barn on the concrete. Of course, there's also bacterial causes of lameness— they can get infections in their hooves from standing on wet or muddy ground too much." He gestured to the barn again and then coughed. "But then also there's things like milk fever, diarrhea, and other reproductive problems that might be a cause of deciding a cow's spent. It can be expensive to treat these kinds of things — especially for a cow that's already almost spent. It might just make more sense to send them on [to slaughter]. Let's move down here to look at the milking string." He led us further down the gravel drive. I scribbled some notes to look up milk fever at home.

Researching later, I learned that milk fever is a condition called hypocalcemia (or low calcium levels in the blood) and is caused by the onset of lactation after birth when high levels of calcium are leached from the body.[9] Hypocalcemia is common due to the physical strain and nutrient depletion caused by repeated impregnation and intensive milking in the industry. Some of these diseases, like bacterial diarrhea, are generally preventable with the use of vaccines and treatable with various antibiotics and other medications, and there is a large market for these within the industry. However, as Homer indicated, depending on the health and age of the cow, or the economic priorities of the farm, it may not make economic sense to treat the condition, in which case they are sent to slaughter when these diseases arise.

"OK, so, can you tell me a little a bit about slaughter? When a cow is deemed 'spent,' then what happens?"

"Sure, yeah. When they're spent, they're sold for slaughter. At Ansel, we work directly with a [slaughter] plant, or we'll sell them at auction. There's some variation in how farms deal with this part of the process. Some will go straight to the plant. Others sell at auction. Sometimes cattle buyers come to the farm to buy them. And then for real small operations, there's mobile slaughter. You heard about that?"

"Yes! Actually, I did my last project on slaughter processes and focused on the mobile slaughter unit as part of that." Mobile slaughter units are USDA-approved slaughterhouses on wheels (in the back of a semitrailer truck) that come to the farm to slaughter animals on very small scales — usually just a few animals at a time.

"Yeah, OK, good. Our dairy cows end up at McDonald's and places like that."

"Oh really? Is that the case for most dairy cows?" I asked.

"Yes. If you're eating a burger from a fast food joint, you're most definitely eating a dairy cow."

"Why's that?"

"They're worn-out—they're not good for much else other than cheap, processed meat or ground beef."

"Interesting." I said aloud. To myself, I thought about how interesting it was that this farm, which sold its dairy products at regional farmers' markets, was also connected directly to industrial fast food meat production. Purchasing food at a farmers' market, it is easy to imagine an undeviating link between consumer and the small farm where it was produced. Yet, even smaller farms, who advertise themselves as alternative producers at farmers' markets, are often entangled in industrial production and consumption practices. Consuming milk produced on a smaller farm may maintain a connection with the fast food industries where those cows' bodies end up as cheap meat.

I also let my mind travel to the cow's worn-out body and the term *spent* in this context. *Spent*, of course, refers to the fact that she is "used up" and no longer reproductively viable, but it also signals her declining economic value. Her body and its productive and reproductive capacities are capital (money to be made) enrolled in the process of further producing goods and profits. When this viability is gone, this particular capital is spent. The farmer makes a careful economic calculation in which the cost of maintaining her is weighed against her profitability as a milk producer. Her body's value to the industry is in rapid decline in its current state, and she is sent to slaughter, whereby the last bit of capital is extracted from her body in the form of meat and other products.

I looked up from my notebook and Homer was looking at me, expectantly. Glancing back down at my notes, I asked: "Sorry, just to loop back, you mentioned artificial insemination. Can you tell me a little about that? Do you keep bulls on the farm at all?"

"We have a Jersey bull on the farm as back up, but we rely mostly on artificial insemination. It's safer. More reliable." Homer stood with his hands in his pockets, the bill of his baseball cap pushed up to reveal his eyes. "Bulls can be dangerous to handle. They can hurt the females when they try to mount them. It's just more trouble than it's worth. We order semen from a supplier—usually ABS or Select Sires—and do

the artificial insemination right here on the farm. We want 'em pregnant at fifteen months for the first time, so we need it to be reliable so
we don't waste any time."

Returning later to agricultural extension materials and semen sales
instructions, I found a more detailed description of the artificial insemination process. Artificial insemination, or AI, is performed using
semen from a bull (the production of which is explored later in the
book) and an insemination gun. The animal is restrained in a darkened
box that holds her in place and blocks her vision (a darkened area with
restricted vision will typically calm a cow) for ease of insemination.
The human worker inserts one hand into the cow's rectum, which enables the inseminator to physically manipulate the reproductive tract
into the desired position, while the other hand wields the insemination gun. The worker inserts the insemination gun into the vagina,
through the cervix, and deposits the semen into the uterus.[10]

"And in terms of size, you mentioned on the phone that Ansel has
five hundred cows? Is this considered a small farm?" I shifted the subject.

"Yes, that's right. Five hundred—although some are over at our
rented acreage. We're on the smaller side, but I would probably call this
midsized. You know, there are dairy farms in California with thousands
of cows, so compared to that, we're pretty small. Really, a farm needs
at least a hundred head [cows] to be sustainable just selling dairy. You
can have less cows if you're also selling other crops or goods, but if
all's you're doing is dairy, I'd say you need at least a hundred to make
it. You can't really stay afloat with less than that. So, yeah, we're definitely smaller, but we use a lot of the same technologies as the larger
farms. You know, artificial insemination, milking machines—that kind
of thing. I'll show you the milking parlor in a minute. Also, the breeds,
too. Since I was growing up, there's been a real shift to Holsteins and
the larger farms are usually almost all Holstein. We've definitely included more Holsteins in our herd here as a result."

"Why's that? Can you tell us about the breeds?"

"Well, you've got lots of different breeds of dairy cows—Guernsey,
Brown Swiss, Ayrshire—but in the US, it's really just Holstein and

Jersey, like we have here, that make up most herds. People like Holsteins because they produce such huge quantities of milk. They take more food to maintain, but you get more milk. Jerseys take less food—they're so much smaller. They're still real good milk producers and they have higher butterfat content in their milk. You'll never get rid of Jerseys because people are used to that taste in cheese."

Indeed, when I later researched in more detail the different breeds, I found that Holsteins, the quintessential US American "dairy cow" with characteristic black-and-white patches, comprise 90 percent of the US dairy herd.[11] Physically, Holsteins are large (around fifteen hundred pounds and fifty-eight inches tall at the shoulder), consume large amounts of feed, and produce high volumes of high-quality milk.[12] While other less common crossbreeds have been bred for qualities like disease resistance, fertility, ease of calving, and strength, Holsteins have been almost exclusively bred for milk production and docility. As a result, Holsteins can produce excessive quantities of milk (upwards of sixty pounds per day) but they frequently suffer from mastitis, mobility issues, and infertility—all problems that cause them to be slaughtered at three to seven years of age. Jersey cows comprise 7 percent of the US dairy herd and are the smallest dairy breed (about a thousand pounds) with a creamy brown coat.[13]

Homer continued: "So, we have some Holsteins here, as you can see, but we mostly stick to Jerseys because of that high butterfat content. The milk they produce is perfect for cheese. And they're more affordable to house and feed."

We stood looking into the pen that contained the milking string. Cows crowded at the fence, reaching through to access the long troughs of feed. One of them had a small new nub of a horn growing on one side of her head; the other side had no horn. I asked Homer about it.

"Oh yeah, well, most of these cows were dehorned or debudded when they were young. I mean, their horns were removed. But sometimes it doesn't take if it's not done quite right and they grow back and you have to remove them again. So here you have one growing back in."

I made a note to look up debudding and dehorning later. What

I learned is that disbudding, as it is more commonly known, is one type of horn removal and involves removing the beginnings of horn buds on very young calves by destroying the horn-producing cells in the forehead with a hot iron or caustic paste. Hot iron disbudding burns the horn-producing cells and is the most common method. Caustic paste chemically destroys the cells. Both methods are painful and done without anesthetic; they can cause long-term tissue damage to the surrounding areas, according to the American Veterinary Medical Association.[14]

Dehorning is a procedure performed on older animals once the horns are fully formed and involves the removal of horns that have already grown in. Common tools used in dehorning are handsaws, obstetrical wire, or a device called a keystone dehorner, which looks like a long-handled set of gardening shears—all of which cause significant bleeding.[15] In both disbudding and dehorning, animals must be tightly restrained because the procedures are so painful that the animals often balk and try to escape.

Throughout the rest of my fieldwork, I saw only a few animals—mostly bulls—with horns. For how few animals turn up at auction or live on farms with horns, you would think they aren't common among bovine animals. It is a common misconception that only bulls have horns. In fact, many dairy-breed animals—both male and female—grow horns, and the industry sees horns as a problem to be managed through removal. The removal of horns is routinely performed on calves usually before one or two months of age. The logic behind removing the horns is to prevent bruising the flesh of the animal (or other animals) prior to sale to become meat, to prevent injury to farmers and other animals, to increase efficiency through faster and easier handling of animals, and to save space (animals with horns require larger feeding troughs and more living space).[16] While horns can be very dangerous for humans handling animals in agricultural spaces and farmworker safety is a real concern, economic factors such as those just mentioned are key motivations for the widespread practice of horn removal. Although dehorning and disbudding pose significant concerns related to animal welfare and animals' experiences of pain on farms (especially so since no pain management practices are used when removing horns), the United States has no regulations for these procedures.

"What do they do all day?" My dad asked, gesturing at the milking string.

"Well, this. And then they're milked three times a day—in the morning, at noon, and in the evening." I looked in across the pen and watched the animals pacing back and forth in the pen, some eating and some drinking water, and they occasionally licked or nuzzled one another. Mostly they stayed outside, but some ambled into the barn, and a few were lying in the barn, in open stalls with concrete flooring. Homer continued, pointing at those who were lying down in the stalls: "They lie down, chew their cud and talk to their neighbors while they wait to be milked."

"Where's the milking parlor?" I asked.

"Just through there." Homer pointed through the barn where the cows were lying down in stalls. "We'll walk around there now." Homer led the way, while continuing to talk. "The milking parlor is attached to the barn so that it's easier and quicker to herd them all in there three times a day. We used to pasture the cows during the day, but as the herd increased, it just got too time-consuming to herd them all the way back here every time it was time for milking. This is easy, and they get a nice feed mixture that gives them plenty of nutrition. Time is money, and with milk prices what they are, we have to get as efficient as we can." Homer paused to reach down to the long trough of feed. One of the cows sniffed his arm and went back to eating as he grabbed a handful of feed to show us.

"This is called total mixture ration, or TMR for short. It's a combo of hay, grain—we use corn and cottonseed—and silage. If you don't know what silage is, I'll tell you." Without waiting for us to respond, he continued: "Silage is a mix of things like grasses and corns (sometimes other grains) that are cut when they're young. They get mixed up as this kind of moist, fermented mixture. It's good for helping the cows digest protein. Corn passes right through cows, so it has to be ground up. This helps with the digestion and it gives them some grasses and other things that make the feed more well-rounded. Nutritionally, I mean." Indeed, corn is not a natural feed for cows, and their bodies have difficulty digesting the grain.[17] You might wonder why farmed animals, like cows, are fed a grain that they cannot digest. The answer is multifaceted: cows gain weight quickly and efficiently when they are fed corn, and corn is a cheap food source because it is heavily subsidized by the US government, which incentivizes its use in all sorts of food production (the documentary *King Corn* is an excellent film on the role of corn in US agriculture, and Michael Pollan's book *The Omnivore's Dilemma* also has a terrific section on corn).

"We grow our own silage. You might have seen that big pile covered with plastic and tires when you came in?" I nodded in agreement, "That's the silage—the plastic helps with the fermenting. We get the other feed from folks who grow it in eastern Washington and Idaho, but we use our acreage here to do our own silage."

Homer led us behind the barn and we came around to the milking parlor. It was part of the barn but had separate entrances where the cows filed in to line up to the milking machines.

"Wow!" my dad exclaimed when we saw the milking machines. "How many can you milk at once?"

"We can do twenty at a time. Ten on each side," Homer replied, as he led us into the large room. "There are lots of different ways that milking machines can be laid out. Some are rows like this with different numbers of stations for cows. Some are laid out in a circle, and some of those circular ones even rotate letting cows on and off as it turns."

At first, looking at the milking machines, it was hard to tell what we were looking at, without the cows there. There were two rows of machines, with cement and metal grating to let liquid through. In between the two rows of machines, the floor dropped down a few steps. Homer pointed to the lower level: "OK, so we stand down there and then it's easy to sterilize and attach the teat cups without having to bend over. The cows stand up on either side in two rows of ten. They're led in, line up, and then we dip their teats in an iodine mixture," Homer said, holding up a large plastic bottle containing the dark, reddish-brown liquid, "and we wipe the teats real firmly to get the milk to come down. Then we can attach the teat cups, which are these things that fit right over the teat." He indicated a cluster of rubber and metal cups attached to a series of hoses that fed into the metal machinery. "Then the machine does all the work. It suctions two teats at a time and the milk goes up through those tubes into the machine. It gets collected in a big tank in the other room. The machine has an auto shut-off, so when it's done it just lets go of the teats, we dip the teats in the iodine again, and then the cows all go out and the next twenty come in. Milking machines make this work so much easier. Most farms use them now—I mean, here, we can milk twenty cows in about five to seven minutes. We could never do that by hand that fast."

The use of milking machines, common even on small farms in the United States since the early 1950s, was familiar from my textual research. Increasingly, more fully automated milking systems are being adopted worldwide to contain the cost of human labor. Purveyors of these milking systems advertise that they increase animal welfare by giving animals a "choice" about whether to be milked and when. In fact, geographers Lewis Holloway and Christopher Bear, who have researched the adoption of these milking technologies, have revealed that, although animals do choose to go into the enclosure where they are milked, they are incentivized to do so (this enclosure is where they are fed) and discouraged from loitering too long (the structure delivers a mild electric shock to encourage the cow to leave).[18]

Homer walked over to some hoses with spray nozzles that were hanging from the ceiling. "After each batch of twenty cows, we spray down and clean the whole room. Then the next twenty come right in."

I looked up and down the length of the room and was struck by how every aspect of the room and machinery was designed for efficiency. This is what industrialization processes look like—a shift to mechanized labor that is driven by a need for the efficient accumulation of capital. And with the very close profit margins of milk production, mechanization of time-consuming processes like milking are necessary for farmers to keep up and stay afloat financially. This thought resonated with Homer's earlier comment about how pasturing the animals took too much time, a reflection of how animals' lives are shaped by economic logics that determine efficiency and profitability in day-to-day raising and handling practices. This logic and shift to mechanization also eliminates much human labor: keeping the cows in pens adjacent to the milking parlor and using machines that can accommodate twenty at a time means that the number of farmworkers needed to manage a herd of five hundred is relatively low. During our tour, we encountered half a dozen Latinx farmworkers doing the daily labor of the farm (feeding, scraping manure, driving pick-ups in and out of the farm), and a small group of employees were working in the cheese production area, making cheese. This labor was gendered and racialized: the farmworkers were all Latinx men and the cheese makers were all Latinx women. The farmers (who own and manage the place) are white, and a couple of white women (family members, I learned) run the cheese shop, selling cheese in a retail context.

Importantly, economic logics of simultaneously increasing production while decreasing the costs of production not only shape animals' lives in the dairy industry, they also govern how human farmers and farmworkers interact with and care for the animals, how precarious farmworker labor conditions and pay are, and how farmers and farmworkers, themselves, survive. The economic pressures placed on dairy farmers are immense: the low prices for milk, paired with the high costs of machinery and inputs to industrialize or streamline production, can make dairy farming an economically risky venture. This, coupled with the growth of nondairy milk alternatives and shifts in consumer trends, puts pressure on dairy farmers to become more efficient or risk going out of business. But this efficiency doesn't always pay off; increasingly, US dairy farmers are producing more milk than there is a market for. In 2016, the *Wall Street Journal* reported that "more than 43 million gallons' worth of milk were dumped in fields, manure lagoons or animal feed, or have been lost on truck routes or discarded at plants in the first eight months of 2016, according to data from the U.S. Department of Agriculture. That is enough milk to fill 66 Olympic swimming pools."[19] This excess only seems to be increasing: in the first five months of 2017, farmers had already dumped seventy-eight million gallons of milk, which is an 86 percent increase from 2016.[20] Overproduction of food commodities is not uncommon in the contemporary industrialized food system; in the case of overproduction of milk, producers and the USDA find ways of trying to create markets for dairy products in order to avoid this kind of waste. For instance, the *Wall Street Journal* also reports that Dairy Management Inc., a dairy industry marketing firm, has been working with McDonald's, Domino's Pizza, and Taco Bell to develop more dairy-intensive menu items in order to increase markets for dairy.[21] The USDA regularly buys significant quantities of dairy from producers to bail out farmers who have excess milk they can't sell; as of 2017, there were over 800 million pounds of US-made cheese and 272 million pounds of butter in reserve.[22]

On the farm, to stay afloat in this kind of market, costs of production (human labor, the most time-consuming practices of animal care, and less productive animals) must be cut. For farmworkers, the result is routinely low wages for dangerous (and sometimes temporary) work; for workers who may be undocumented, the precariousness of their employment and immigration status makes organizing for better working conditions a dangerous proposition. This political economy of agricultural production, then, puts unique pressures on each of the lives embedded in this system. The pressures on farmers' and farmworkers' lives and labors, in turn, fall on the lives and labors of the cows on whom the industry relies. Talking to Homer made me keenly aware not only of the existence of these pressures but also of how human farmers', farmworkers', and animal laborers' futures are entangled.

Homer led us out to the other side of the barn where the silage pile loomed. As we exited the milking parlor, we encountered a man driving a tractor with a scraper attached to the front that dragged along the ground and scraped manure into slits in the floor of the barn.

"What happens to that liquid down there under the barn?" my dad asked.

"Well, the liquid manure gets swept into a holding pit. Then we come in later and pump it out into a truck and take it out to the manure lagoon. Later on, we can use that to fertilize the silage crops." He pointed to the back end of the property. Manure lagoons are the most common way of dealing with waste from animal agricultural production facilities. These lagoons, which are earthen pits in the ground, hold the manure and then, often, the liquid is collected and sprayed over fields as fertilizer. Manure lagoons have been linked to a number of deleterious environmental and health impacts for areas surrounding them, and human workers have been known to drown in them when they have become overcome by the stench and fallen in. In a process called environmental racism, communities of color and economically marginalized communities are disproportionately exposed to these kinds of environmental hazards because farms — especially large-scale industrial farms — are sited in close proximity to these communities. As a result, air, water, and soil pollution shape their health and resilience.

Homer explained that "inspectors fly over the area to check on the lagoons and make sure they're not causing spills or overflows. Farms can get fined for that, if they're not taking good care of the lagoons. Flooding, too, can be a problem. If we get a lot of rain and there's already a full pit, that can cause an overflow and then the manure seeps out. The state and county worry about it getting into streams and rivers. So, yeah, there can be fines for that."

Later, as I was writing up the research for this book, I continued to read about the impacts of storing manure in this way. Manure lagoon spills and leakages are not uncommon in dairy and other animal agriculture industries. In fall of 2013, Pomeroy Dairy in western Washington was fined $6,000 for a manure spill at their farm that violated Washington's Water Pollution Control Act. The farm did not report the spill and only admitted to the spill occurring when confronted by the Washington State Department of Agriculture inspector who was called after local public works officials found extremely high levels of fecal coliform in nearby waterways. The contamination spread as far as five miles downstream, killing fish and other aquatic life and posing a serious risk to human health. In 2015, Snydar Farm in Whatcom County, Washington (the county with some of the highest levels of contaminated drinking water in the state), was also fined $12,000 for

ongoing manure-based contamination of local waterways; this farm
has been the subject of ongoing investigations by the Washington State
Department of Ecology and has, in fact, caused continuously increasing levels of contamination between 2010 and 2015.[23]

Water contamination also reaches beyond local drinking water issues. Runoff from farms and manure lagoons is also a major contrib-
utor to "dead zones" in oceans and lakes, a phenomenon where heavy
pollution reduces oxygen in the water to the point where most marine life cannot be supported. Dead zones are increasing around the
globe, causing swaths of oceans, lakes, and other bodies of water to
be uninhabitable to most aquatic life. The Gulf of Mexico has a massive dead zone as a result of agricultural runoff (of manure, pesticides,
and herbicides) flowing downstream from various waterways to the
Mississippi River and then out into the Gulf. Interestingly, dead zones
are not completely dead—indeed, as geographer Elizabeth Johnson
argues, they are teeming with life: algae, jellyfish, and other living or-
ganisms can survive and flourish in marine worlds that are otherwise
deadly to most creatures.[24] But these so-called dead zones dramatically
reduce biodiversity by killing off other lifeforms and create conditions
where phenomena like jellyfish blooms make human use of the water
(for swimming, fishing, etc.) difficult or impossible.

As I stood there, watching the manure scraper do its work, I thought
about the cumulative effects on the environment of farming animals.
This was just one relatively small farm. There were farms like this—
and much bigger ones—scattered all over the state of Washington,
up and down the West Coast, across the United States, and beyond.
Industrial farming practices are being exported around the world to
countries interested in producing animal products for domestic consumption and for export economies as part of global economic development projects. At that moment, watching that singular manure
scraper, I felt a sense of the impact—the overwhelming scale—of animal agriculture in a way I hadn't considered in such an embodied way
before. The growth and expansion of animal agriculture globally continues each year. In fact, since I visited Ansel Farm in 2012, they have
doubled the number of cows they raise—to nearly a thousand.

Homer glanced over at me as I was rubbing my nose with the back of my hand. The smell of rancid manure was very strong and burned the inside of my nose. The liquid manure sloshed through the slot in the ground, pouring into the pit like a chunky brown waterfall, and I felt sick.

"Powerful, huh?" Homer chuckled. "You get used to it. That there's the smell of money."

This was a line common in the industry—I heard this repeated several times throughout my research and I read other work that quoted this line. Manure: *the smell of money*. If this was the smell of money, then was environmental degradation—manure lagoons, polluted waterways—just a necessary part of the capital accumulation process? I couldn't help but think that this level of pollution might be just one of many costs of commodifying life. But I also thought about the costs of commodity production for the animals themselves. What was their experience of this accumulation of capital? What bodily impacts did they experience so that profit could be made from selling dairy products?

Being at Ansel Farm was an illuminating experience in this regard. Although the farm was structured around the reproductive functions of the cows who lived and labored there, and Homer clearly cared very much for the cows there, the organizing logic of the farm made it so that the cows themselves were abstracted. Their days and lives were organized around the production of commodities: three times a day they were herded into the milking parlor to be milked, while in between milkings they waited in the adjacent pen, pacing, eating, and drinking, crowded in against their neighbors. The cows were integral to the commodity production, but as singular beings with interests of their own, they were less visible. Even the ones who stood out, like the heifer with ear tag #6490, were framed through their commodity potential: her impending calving, her future as a milk producer, the longevity of her fertility, her eventual decline and slaughter for meat. This dominant logic at the farm was powerfully contrasted with the moments where Homer expressed a deep sense of care for the animals: when he mentioned the traumatic dimension of removing calves shortly after birth, or the warm and loving way he talked about and looked at the

Jersey cows. He was not without a significant sense of responsibility and care for the animals themselves. But these feelings seemed to be at odds with the structural context in which Homer, this farm, and these cows existed. To linger too long on the emotional experiences of the cows troubles what is required in commodifying them: a way of the seeing the animals as producers of commodities needed for the farm's survival, as well as eventually commodities themselves. He was quick to move on from those moments and to re-center his commentary on the process of milk production and the productive capacities of the animals. It struck me that keeping these feelings of emotional connection at bay may be a necessary survival mechanism for farmers who may care deeply about the animals they farm and who struggle emotionally with certain routine realities of dairy production (like calf removal after birth, or slaughter).

It also struck me that, while it is, perhaps, a necessary part of the production process, this level of abstraction and commodification also involves a certain level of mundane violence in the monotony of a life completely structured around the ease and efficiency of being milked. The cows' lives at Ansel are configured to accommodate the demands of a human industry in which they serve as milk machines, by which I mean they are made the means of producing a marketable commodity. The cows themselves—by which I mean their lives, their upkeep, their health—represent a capital investment with a market value, and, thus, the cows themselves become commodities with a market value. It was at Ansel that I first noted the everyday violence of this process when the commodity is a living being. The dimensions of dairy production that animals experience on a daily basis are, indeed, driven by economic interests. And I want to be careful to highlight that it's not that farmers are uncaring or don't have deep feelings of love and care for the animals they farm. But practices now common in the industry (and taken up even by small farmers) must involve careful calculations of cost and profitability and illustrate the need to eke out the most capital from each body in order to survive in an increasingly competitive market. And these modes of commodification are explicitly gendered.

Bodies determined to be "female" by the industry during the sexing process at birth—the heifer and the cow—routinely have their reproductive systems appropriated for the production of milk as a commodity.[25] This reproductive appropriation spans beyond the dairy industry—it occurs across the institution of animal agriculture; female animals are bred, labor, and die for their reproductive capabilities. Hens raised for eggs are some of the most intensively used animals in the world: 95 percent of egg-laying hens in the United States are housed in battery cages too small for them to spread their wings, and most hens are debeaked to prevent cannibalism in such close quarters.[26] Hens are slaughtered at two to three years of age, when they are deemed "spent."

Sows, who are judged to be highly intelligent and social animals by human standards of intelligence and sociability, are repeatedly impregnated and housed in gestation and farrowing crates too small to turn around in order to breed generations of pigs slaughtered for pork.[27] Sows are impregnated as soon as possible after giving birth and will typically birth at least two litters of piglets per year; when a sow's reproductivity declines, she, too, is slaughtered for meat. Outside the United States, models of intensification in animal agriculture are taking hold as chickens, pigs, and cows are being raised intensively around the world. As anthropologist María Elena García has highlighted in her research, these models are also taken up in practices of farming other nonhuman animal species, like guinea pigs, who are increasingly bred intensively and slaughtered for meat in Peru and other Andean regions. García explains how industrialized chicken-raising practices have been adopted and intensified in guinea pig production and emphasizes specifically how the female guinea pig and her reproductive potential is central to this process.[28]

If the heifer with ear tag #6490 at Ansel Farm continued on to a routine life trajectory for cows in the dairy industry, her life will have been better than most cows used for dairy production: the cows at Ansel Farm were in better condition than any other animals in spaces of commodity production that I encountered in the rest of my research. She will have waited in the maternity pen to give birth to her first calf, an experience that will have started her on a life-long cycle of waiting in pens to be milked and impregnated and to give birth to keep the cycle going, a process that inevitably leads to declining fertility and decreasing productivity. Each year, Homer will have made an increasingly careful calculation to determine whether it was profitable to impregnate her again. After several years, likely even by the time this book is published, the answer will have been no, and she will have been sold at auction for slaughter.

That somber reality nagged at the back of my mind as I thanked Homer for taking the time to show us around and for sharing his knowledge of dairy production with us. It nagged at me as I patted the heifer with ear tag #6490 and scratched behind her ear once more, in parting. As we drove away from the farm and out onto the country highway that led to the auction yard, my dad and I chatted about the farm—filling out my notes with more detail based on what we both could remember and scribbling questions in the margins of my notebook to be answered through further research later on.

As I was putting the finishing touches on this manuscript in summer of 2017, I contacted Ansel Farm and asked to speak to Homer. I wanted to share with him what I had written about the farm before it was published. I was informed that Homer no longer worked at the farm. Chatting with the person who answered the phone for a couple of minutes, I learned that shortly after I had toured the farm in 2012, they had discontinued farm tours for "health and safety" reasons.

4

LIFE FOR SALE

Although auction yards are public spaces—events that anyone can attend—I was still anxious that we would be turned away. After having been denied by so many farmers, to come to a place where these farmers commonly mingle and sell their animals was intimidating.

Before heading into the auction house, we wandered around to the side of the parking lot where an expanse of pens and chutes stretched out behind the auction building, which looked almost like a large white farmhouse with barn-red trim, the paint peeling and faded. There were groups of cows, heifers, and steers enclosed in these outdoor pens, some just a few to a pen and some pens so crowded that the animals were climbing on top of one another in an effort to see out. Still other pens were empty and a lone rooster pecked the ground in one of them. My dad walked right up to the fence and crooned at the waiting animals. Some of them turned to look at us, and my dad snapped some photos with his camera. I scribbled notes about the crowding in a number of pens and the condition of the animals.

Conditions that cause cows, bulls, steers, and calves stress are increasingly well-documented; academic researcher E. M. C. Terlouw and her colleagues point out that these stressors include new situations, handling and loading, disruptions of social groups, fatigue and exhaustion, and, sometimes, encountering unfamiliar animals.[1] All of these conditions are present at auction yards. Leaving the farm, animals are handled and loaded onto transport trucks. They are separated from their social groups (and this disruption occurs for those animals who are sent to auction as well as those who stay behind at the farm).

On arrival at the auction yard, they are unloaded into a wholly new situation, and they not only encounter new and unfamiliar animals but are also housed in close proximity to them. They are then handled intensively, exposed first to the new surroundings of the holding pens, followed by the auction ring itself, and then other new situations, loading and unloading, and encountering new animals as they are transported to their next destination (the dairy farm, the feedlot, the slaughterhouse). This process and the stress it causes leads to fatigue, another known stressor of cows, which in turn causes additional stress. As we watched the cows and steers waiting in the pens outside the auction buildings, many of their eyes bulged in fear, the whites showing, and saliva foamed at their mouths. They were covered in feces and caked with mud. Bellowing in the pens echoed across the auction yard.

My dad put away his camera and I tucked my notebook in my bag as we entered the auction house. I stopped in the office where buyers complete transactions involving the sale and purchase of farmed animals to ask about how the auction worked. I told the woman sitting at the front desk that this was my first time at the auction, and she told me that I should register with her if I was planning to bid on any animals. I told her I didn't plan on bidding—that I was here just to watch— and she pointed me in the direction of the auction hall. I thanked her and walked down the hallway to the room where the animals were auctioned. On the walls of the hallway were posters advertising animal products, a few with the familiar "Beef: it's what's for dinner!" tagline and a plate showing a giant steak or cheeseburger. Across the hall from the office was a restaurant that served burgers, fries, and pie.

We made our way into the auction hall and found a seat in the bleachers. We were a few minutes early, and there were not many people in the auction hall—many were still out back, looking at the animals before the auction began and asking questions of the sellers and auction employees.

While we waited, I noted the spatial arrangement of the auction yard and I later saw this design echoed in other auctions I visited. Farmed animal auctions are usually held in large buildings designed for functionality. Directly connected to the auction house are a series of pens and chutes for containing the animals while they wait before or after the sale. At some auction yards, animals are unloaded at one end of the yard before the sale and loaded onto transport trailers at the other end of the yard, creating a flow of animals through the auction yard, by way of the auction ring. At others, the animals are unloaded and loaded in the same area, creating a flow of animals back and forth, in and out of the auction yard. I scribbled messy sketches in my notebook, trying to capture these spatial logistics and, later, the movement of animal bodies through this everyday space of buying and selling.

As the auction starting time neared, more people—mostly older white men—sauntered into the room and found seats in the bleachers. A man with his young son and daughter came in and sat next to us, and the little girl—maybe six years old—turned on the bench to tell me that she was excited to see the cows. She held up a picture book with a smiling black-and-white cartoon cow on the front and declared, "I'm going to see cows! Just like the one in my book!"

"Oh yeah? Me too. This is my first time here. Have you been here before?" I asked.

"No. I haven't been before. We're going to see cows and . . ." Her voice trailed off as she got distracted by a dog walking at the heel of one of the buyers over to the bleachers on the far side of the hall.

The auction start time came and went and there was no sign of the sale beginning. We waited for another twenty minutes or so. I saw a group of employees and auction attendees chatting off to the side and, when a woman who had been talking with the auctioneer came and sat in the bleachers near us, I asked her about the delay.

"Oh, a steer jumped the fence and took off running down the highway. They grabbed their rifles and had to chase him down in their trucks."

"Were they able to capture him and bring him back?" I asked.

"Nah, they cornered him a ways down the road and shot him." Her voice was so relaxed, and she stated this fact so casually, that the steer's death came as quite a shock and before I knew it, my eyes welled up with tears and I looked away from the woman to hide my grief. Seeing my response, she said, "Yeah, it happens now and then. Too bad, too, because that steer woulda made some high-quality beef."

Later, I thought about this steer's escape as a form of resistance to the conditions that work to commodify him. Jason Hribal, in his book *Fear of the Animal Planet: The Hidden History of Animal Resistance*, writes about animal resistance in the contexts of zoos, aquaria, and other spaces of animal entertainment, arguing that while these acts are often reported as isolated incidents or as "good animals gone bad," they can, in fact, be understood as a continuing history of animal resistance struggles. I have written elsewhere about news stories and personal anecdotes of individual farmed animals resisting their captivity and the responses of the public to these acts: some animals who escape are viewed as deserving of a different life by the public, rescue groups, and animal control officers involved in their capture.[2] Others, like the steer that jumped the fence outside the auction, are killed.

A few minutes later, the auction began. I felt my body tense as the door opened and the first cow came through into the ring. She was a large Holstein, more worn-looking than the cows at Ansel Farm, but the auctioneer's comments revealed that she was still viable as a producer of milk and new calves. "She's calved once already—she's good stock. Lotsa milk outta this one! Just look at those udders. Ooh-wee!" The auctioneer began calling the escalating prices in a hypnotic hum. Immediately I was lost, trying to understand the structure of the pricing, only occasionally catching the subtle nod of an audience member gesturing his bid. In under a minute, the bidding was over, and the two teenagers working the ring herded the cow out the exit door. Their job was to swat the animals in the ring with aluminum rods to keep them turning around and visible to the audience and then to herd them out the exit door as efficiently as possible when the bidding had ended.

Immediately, another animal was herded into the auction ring just as the exit door slammed shut behind the first cow. This was a pregnant Jersey heifer, much like the one with ear tag #6490 I had met at Ansel Farm. The woman who had informed us about the steer at the beginning of the auction leaned over and asked, "This your first auction?"

I don't know why she asked—maybe it was the way we looked out of place or maybe it was the look on my face, watching the animals pass through the ring. But I was relieved that she was talking to me and that she was friendly.

"Yeah, it is. It's confusing!" I confessed with a half-smile.

She explained that "pregnant animals go for different amounts based on how far along they are. Usually, they're more expensive the closer they are to giving birth." Because cows are conceptualized as commodities largely through their reproductive value, a pregnant cow's value can be higher than a cow without a calf in utero. Her impending calf will be valuable either as a dairy producer (if she is born female) or, less valuable, for veal or beef (if the calf is born male). Her pregnancy is also proof that she is fertile and can carry a calf. This commodity logic was reflected in the auction pricing and the sales process at the auction yard.

The next sale was a Guernsey-breed cow and calf pair. The cow had resisted, kicking and blocking the man herding her, and there was a slight delay as she and her calf were herded into the ring. Their coats were thicker than the other animals, giving them a fuzzy appearance, and they were orange-brown with white patches. The calf stayed close to the cow, hurrying to keep up with the cow's agitated trotting around the pen. The cow's eyes bulged and she let out loud snorts as she looked for an exit.

The woman next to me leaned over: "Oh man, I was going to bid on this pair, but the way that cow's resisting, it's clear she's psychotic. . . . Too bad, too, because they're real beauties." And they *were* beautiful, the two of them together—gorgeous animals who were clearly closely bonded and in good physical condition. They sold together to a man on the other side of the bleachers. As I watched the cow and calf trot through the exit door together, I contemplated the woman's comment that the cow was psychotic. From my perspective, she had been trying to protect her calf, which seemed to be a reasonable (and not neuro-atypical) response to the unfamiliar and likely frightening conditions of the auction yard. I continued to chat with the woman in the bleachers as we watched animal after animal move through the auction space, their ownership changing hands with the slight nod of a head from the audience and a scribble on the sales ledger by the woman seated next to the auctioneer.

The layout of the auction hall—and this was the same at all the different auction yards I visited—is simple: a large ring with steel fencing separating the audience from the animals. The auctioneer's platform sits opposite the audience inside the auction ring and the auctioneer (always a man at every auction I attended) sits behind an elevated counter with a microphone, calling out the prices and bids. Above him hangs an electronic scoreboard that calculates the weight and price of the animal and displays those figures for all to see. To the right of the auctioneer (i.e., his left) there is an entrance door, wide enough for several large farmed animals to pass through at once and, to the left of the auctioneer, an exit door. Outside of the exit, there is a large scale on the ground. As the animals leave the auction ring, they step onto the scale and their weight is transmitted to the screen above the auctioneer's platform.

Farmed animals at auctions are marked with neon-colored symbols to indicate certain characteristics—whether an animal is sterile, whether a cow is pregnant, and so forth. Animals come to auction in all conditions, and their value is set accordingly. Occasionally, animals are sold in a group, in which case the auctioneer will note that the buyer must purchase the whole lot at the agreed-on price. More often, though, the animals are sold individually. In both cases, they are priced either "per head" (the whole animal) or "per pound" (usually per hundred pounds of weight for cows). Cows sold by the head are usually being sold back into dairy production and their bodies are priced as the whole reproductive unit: a singular entity. In contrast, cows sold by the pound are typically headed to slaughter, their bodies already conceptualized as meat and calculated for the profit that can be gained from the disassembly of the worn-out reproductive body. This was the "dairy market" sale, so animals sold at this auction were intended to be sold back into dairy production. The animals sold at this auction were not yet at the end of their reproductive journey.

Auctions are places of both production and consumption. Animals come in as producers of commodity goods: either for their breeding and/or milking capacities or for their flesh as producers of meat. Auction yards are necessary stops along the production process: they facilitate the trade of animals when a particular producer has no use for them anymore and the sale facilitates the reenrollment of the animal in the production process. But auctions are also places of multilayered consumption of animals. Buyers are obviously involved in the economic consumption of live animals during the auction process when they make a purchase. Auctions are also places where the consumption of animal products is visible as auction workers and attendees wear leather boots and belts and as they eat animals' flesh and reproductive outputs. Like this one, many auction yards have restaurants that serve traditional "American" fare: burgers, fries, pie, ribs, steak, biscuits and gravy, et cetera. At one Washington auction yard, the restaurant is called The Branding Iron, a direct reference to the branding process and a declaration of ownership and appropriation of animals that is reenacted in the consumption of animal bodies at meal time.

Throughout the auctions, I saw hundreds of different brands on the hips of cows—scars burned permanently into their flesh and healed, a mark of ownership. The most common method of branding is hot-iron branding without anesthetic and has been performed for thousands of years.[3] Freeze branding is another method gaining popularity, and tattooing is another method of permanently marking ownership of the animal. Brands are unique to each farm and are a source of pride and history for intergenerational farmers. There are currently alternative (albeit less permanent) methods of identification that are less painful for the animal (e.g., paint marks, microchips, electronic leg bands, and electronic collars), but the continued practice of branding represents a nostalgia for the past and for tradition that is resistant to evolving to meet ethical advances in what is known about animals and the effects of humans' treatment of them.

At auctions, buyers arrive early to look at the animals and decide who they will purchase and then will typically sit down for a meal with friends in the auction restaurant. The interiors of auction halls are wall-papered with advertisements for animal products, like the "Beef, it's what's for dinner" advertisement I saw on my way to the auction hall from the office, or there will be ads promoting the health benefits of milk. These encourage the consumption of animals as a way to support local farms and communities. For many, the consumption of animal bodies (both as living capital to buy and sell and as food) is integral to the auction experience—both as a social activity that supports the surrounding community and as a way to reinforce the way commodifying farmed animals is made routine.

"Dairy cows" are commodities because the contemporary United States is a consumer culture. From the day they are born into the dairy industry, cows are subjected to continual consumption. The productive life of a cow is commodified, expropriated from her, and put to use for economic gain in the production of milk and its various derivative consumables (e.g., cream, cheese, curds, and whey), and in the (re)production of the cow for the next generation of dairy producers. And, once her productivity declines and she is declared "used up" or "spent," she is sold again and sent to slaughter where the cow herself vanishes, commodified yet again by a reconfiguration of her body as meat. And, thereafter, once again—as trimmings, bones, and offal—remnants of her are sold to rendering for the production of an array of other consumer products from leather and soap to bone meal. In the total commodification of the "dairy cow," the animal herself is thoroughly transformed, first as a living being and then as a carcass, into an array of new commodities. A closer look at the word *commodity* reveals the fundamental purpose of commodification; both words are cognate with *commodious* from the Latin word meaning *convenient,* which, of course, means convenient *for us.* A commodity can be sold and is bought because it is a convenience. For the sellers, it is a convenient way of making a profit; and for the ultimate buyers, the consumers, the thoroughly commodified cow used for dairy is a convenient preparation for consumption of government certified nutrients (meat and milk), as well as leather, cosmetics, and fertilizer.

The linchpin in this whole process is the dairy auction. As the auction progressed, I was surprised by the simultaneous efficiency of the sale of animals and the casual attitude of the auction employees and the audience—these sales were clearly a mundane, unremarkable feature of animal agriculture. Animals' circulation as commodities at this auction was just a routine dimension of their ongoing life as dairy producers, and a certain level of efficiency was needed to handle the great number of animals waiting to be auctioned. The bidding for each animal was generally finished in less than a minute, resulting in somewhere between fifty and sixty sales per hour. As I was researching the history of auctions, I discovered that this speed of sale has not actually increased since the 1960s, when fifty sales per hour (on average) were made in "livestock" auctions in the United States.[4] This stagnation in the speed with which animals are moved through the auction, in contrast to the extreme intensification of other aspects of animal agriculture since the 1960s, is likely due to the fact that, according to Ralph Cassady, "the time required to move animals through an auction slows down the operation, and the buyers, some of them ranchers to whom time is a more flexible element, tend to be more casual in their bidding."[5]

Geographically, the auction yard is a place noteworthy for the tensions between the necessity of containing the animals and the need to facilitate their controlled movement through auction spaces (see figs. 4.1–4.4). For auctions to facilitate the efficient sale of animals, it is crucial for them to move smoothly through the network of pens and chutes and to move quickly in and out of the auction ring during the sale. Auctions are designed with this in mind. Nearly every segment of fencing is hinged and moveable, so that together the fencing can be easily arranged by one or two auction workers to create a place of containment (a pen) or a place of movement (a chute). Auction workers are responsible for managing how these areas are spatially ordered. In other words, they determine which segments of fencing are involved in containing the animals and which are not. With large animals like cows, steers, and bulls, an auction worker's memory and awareness of the current layout of the auction yard is critical for worker and animal safety. An error in judgment that leads to a frightened twelve hundred–pound cow barreling down a chute that a worker thought was a contained pen could lead to the injury or death of workers or other animals. Similarly, a missing segment of fence in the wrong place could

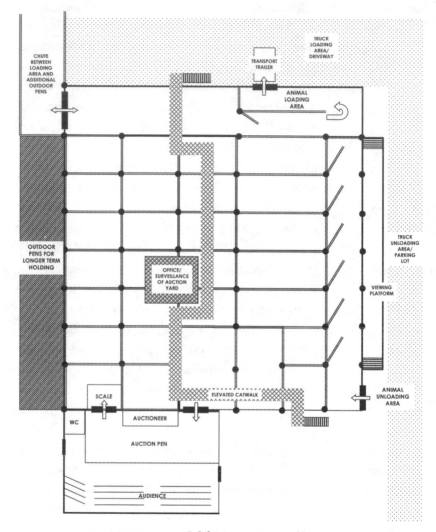

FIGURE 4.1 California auction yard layout

lead to animals escaping. And so, the management and knowledge of
each segment of fencing is important for the efficient containment and
movement of the animals leading up to, during, and after sale.

Tools are also integral to the bodily management of animals as they
move through the auction yard. On arriving at the auction, each animal receives an identifying sticker with auction number and barcode.
This is used in tracking and selling the animals during their time at the

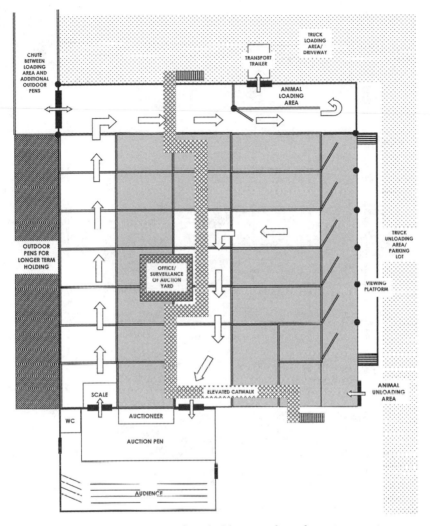

FIGURE 4.2 Movement from holding pen through auction to exit

auction yard. A rod or paddle is common for driving animals through
the chutes and urging them to turn around in circles in the auction
ring. The rod is a lightweight metal or plastic rod that workers use to
slap the rump, side, or face of the animals to keep them moving. The
paddle, which is broad and made of plastic, is attached to the end of a
rod; these usually have rattles inside to startle the animals into moving forward. An electric prod (a wand that shocks the animals with an

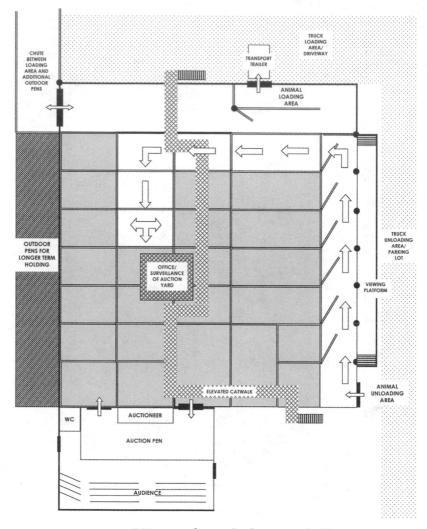

FIGURE 4.3 Movement from unloading area to holding pens

electric current) is often used to drive animals through the chutes and into the transport trailers. Interestingly, I never saw the electric prod used in the auction ring in front of the audience, but on a number of occasions, I did see it used heavily in the rear holding pens and chutes. These technologies are used in the routine movement of animals through the auction yard. When an animal does not move through the space in a usual way—in other words, when an animal collapses

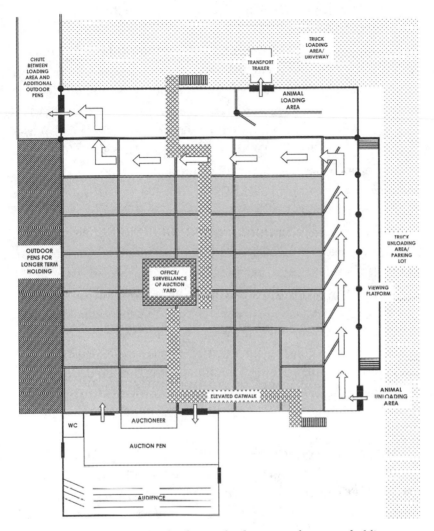

FIGURE 4.4 Movement from unloading area to long-term holding

or escapes—these devices are used in a different way. In the case of a
collapsed animal, the electric prod or some other device that causes
pain will be used to try to force the animal up. In a case where the animal cannot rise, they will likely be shot. Firearms are used in extreme
circumstances at the auction yard, either to shoot a nonambulatory
animal who is unable to get up or to shoot an animal who escapes, like
the steer at the beginning of this auction.

Later that day, another cow-calf pair came into the ring. There was a delay in the auction, and our attention was drawn to the rear holding pens. We could hear the sound of hooves on the wooden ramp leading up to the auction pen, the shouts of auction workers, the bellowing of an adult cow, and the higher-pitched calls of a calf. The large auction ring door opened to reveal the back pens and chutes as an auction worker struck a Holstein cow in the face with a rod he was holding while yelling loudly. The cow still would not enter the auction ring; her calf was behind her on the ramp and she refused to leave him behind. The cow and calf were to be auctioned separately, and the auction workers intended to auction one at a time. In the end, they herded both the cow and calf into the ring to avoid any further delay and the auctioneer made an announcement that they would be auctioned separately: the calf first, and then the cow.

The calf was very young—not more than a month old, the auctioneer said, and he sold immediately for $135. The teenagers in the ring worked together, one opening the exit door just wide enough to allow the calf to exit, while the other distracted the cow. The calf, startled, trotted through the opening, stumbling at the threshold. The worker standing at the door smacked his rump with the rod he was holding and the calf leapt forward out the door. The door closed and the calf was gone.

The cow trotted in circles in the ring and bellowed. From the pens behind the auction ring, the calf called back. After the cow sold for $1,600, they herded her out the door, which was then closed, and she was gone. But we could hear the cow and calf continue calling to each other from their separate pens in the rear holding area.

As I watched the sale of the cow and calf unfold, I thought, *this is what it means to commodify a life*. The cow and calf's bond and the violence of their separation was tangible in their calls to each other across the auction yard. At the beginning of this project, I wanted to know: What was involved in commodifying a living body? And in the auction yard, in particular, I was interested in what it means to put a life up for sale. This was one intimate effect of appropriating, buying, and selling life: the severing of emotional and physical bonds within nonhuman family structures.[6] But before I could think too much about this in the moment, the next animal was coming into the ring.

The auction has an unsettling way of revealing that the sale of the animals—the sale of living breathing beings—is at once momentous and utterly mundane. As we sat in the bleachers and continued to watch as a steady stream of individual animals passed through the ring and were sold, I felt keenly aware that something troubling and important was happening here, something that might—fifty, one hundred, two hundred years from now—seem utterly unfathomable. The routine sale of these lively and emotive creatures with deep emotional lives and close bonds with their families and social networks for food production—to be bred, milked, killed, and cooked—involves an almost incomprehensible level of controlled violence. It is an appropriation of animal life that has been thoroughly normalized because of its convenience to us. And of course, after ten thousand years of animal agriculture, it has become a smoothly running process.

The steer shot down on the road; the cow and calf sold separately, their last moments together a frenzied scene of panic in the auction ring; the cows who trotted through the ring and sold without incident—these encounters felt to me like moments that needed to be remembered. They were important junctures in the life courses of these animals, as a sale that took only a minute would determine their future trajectory in profound ways. I imagined an abstract visual representation of this auction yard, where a constant flow of animals would stream into the auction house and then, after the sale, would disperse in a hundred random directions. This level of commodification and movement was overwhelming to think about, especially when scaled up to all the auction yards in Washington, on the West Coast, in the United States, and beyond. Similar to the moment at Ansel Farm when I watched the manure scraper do its work and was struck by the global environmental impacts of farming animals, I was floored in the auction yard by the sheer scale of animal agriculture, the scale of violence through bodily control, separation, and commodity production that millions of cows undergo for dairy production every day. As a way to honor each cow, and her embodied experience of being so thoroughly commodified, I thought I should try to remember each face as she passed through the ring.

But of course, that is impossible, for even as the animals are (usually) sold as individuals, the auction has a profoundly abstracting effect. The mundane nature of the auction yard and the relentless, routine nature of commodifying the animal body insists on the abstraction from the singular animal. Although each animal is sold for her unique re/productive potential (as a producer of milk, calves, and meat), that potential is realized as market value and thus quantifiable as a price. Moreover, she is also just one unit of production in the political economy of global food production. The auction yard lays bare this reality in the efficient nature of her recommodification as she is sold on to her next site of re/productive labor. In the auction yard, the *cow* easily blurs into a stream of *cows*, and the routine circulation of living capital becomes monotonous.

It feels shameful to say it considering the auction's weighty role in shaping the cow's future, but sitting in the auction as a spectator can be incredibly *boring*. This is probably not the case for those who were there to purchase an animal or for those for whom the auction was a social event, but for me — a stranger in this community, there as a mere observer — the auctions were often boring. And this effect of being boring speaks to the power of how the auction abstracts and blurs the singular lives passing through the space. Animal after animal sold without fanfare. They were poked and prodded to pace around on display, they were bid on and weighed, and then they were gone — out the exit door — to wait for transport to their next destination.

Even as I tried hard to remember each face and to note the features of each individual in my notebook (*Holstein cow, 3 years old, slight limp, docked tail, fair condition*), when I looked back through my notes later that day or week, or six months later, I could not of course remember them. They became generic descriptions of any cow at any auction. The ones I did remember were the ones who in some way broke the routine monotony of the sale — the ones who died, the ones whose conditions moved me to tears, the ones who did not move easily through the auction space. The others were regrettably and easily forgotten, and that maybe haunts me more than the few whose faces I do remember.

Throughout my research process, I returned to many auctions, and I share more moments from these various spaces throughout the remainder of the book. I kept returning to auctions for the very practical reason that they are open to the public and an easily accessible site to view a node in the commodity circuit of animal agriculture. I was especially grateful for this access considering the difficulty or impossibility of accessing farms, slaughterhouses, rendering plants, and other sites related to the production of dairy. But there was something else about the auction that haunted me and caused me to return again and again, each time understanding a little bit more about the embodied experience of animals within a global network of dairy production.

What is it about the auction that is so noteworthy? The farmed animal auction is a place where the commodification of the animal life and body is visible in stark detail. This is, in part, because it is the site where animals are bought and sold and are assigned a market-determined value as a commodity. Animals in agriculture are commodified in all kinds of ways and in different spaces. On the farm, their milk is extracted and sold; their calves are sold shortly after birth, either to be raised as calf or milk producers or to be killed, their bodies to be dismembered at the slaughterhouse, a disassembling into parts that transforms the bodies into meat, to be packaged and sold. Farmed animals' lives and deaths are commodified at nearly every juncture. And there is a certain crudeness about the auction—about the routine sale of life in this everyday space—that reveals the mundane violence of commodifying living beings. I rarely saw what would be considered by the law to be cases of cruelty or abuse of animals in the auctions I visited. Instead, what may have been even more troubling were the ways that the everyday practices of industry enact such routine violence on the animal body.

To call these routine practices of the industry "violence" may seem extreme—they are certainly not seen as such within the industry. The auction itself is often a jovial event where farming communities come together and socialize, sharing stories and jokes. But the auction and other routine practices involved in dairy production—like those described in the previous chapter—involve a violation both of the animal's body through appropriating it for food production and of her ability to have a life of her own. The cow's interests are subordinated to the objectifying logic that makes it possible to buy and sell her in the first place. Her interests in raising her calves, having pasture to roam, living in community with her broader social networks, avoiding intrusive procedures like artificial insemination and milking, and having bodily liberty and autonomy—these interests are swept aside, overlooked in the process of transforming her into milk and meat. There is a certain level of violence operating at the root of commodifying life—the process that requires, even in the most thoughtful of farming settings, like Ansel Farm, a prioritization of the animal-as-commodity above the animal as a singular being with a life, interests, and attachments of her own for which there can be no provision in determining her commodity value. As a commodity, the life interests of a cow, those things of value to the cow herself, are eclipsed by a number representing either an investment or a return, a calculation of convenience to humans.

Although this process and its effects are intensified in larger-scale, more industrialized dairy production, these fundamental modes of violating the cow exist in small-scale production as well. This is one reason I have chosen to focus on this particular Washington State auction as the subject of this chapter. Most of the animals passing through this auction yard were coming from smaller farms where the farmers would bring just a few animals to sell. Some of the animals were sold as "organic" producers, having been fed organic feed during their lives. By contrast, an auction yard I visited in California's Central Valley was a site of exchange for much larger dairies. Sellers would bring full-sized truck trailers holding dozens of cows at a time to sell at auction. What I want to make clear as a result of the research I did for this book is that there is a certain level of violence in the mundane, everyday nature of the industry—and while this violence may not be as readily visible, it is insidious, deeply rooted, and continually reproduced.

As my dad and I left the auction yard that day, we watched as buyers loaded their new purchases into transport trailers. These animals
were on their way to new sites of commodification, where they would
be bred, milked, and bred again. Eventually, they would end up back
at the auction yard for a different kind of sale: the cull market auction.
I didn't know what the cull market auction was like that day as I stood
there watching these animals filing into the trailers. This had been my
first auction, and as I would discover later, this sale was the most mundane I would visit; the animals were in relatively good condition compared to every other auction I would see.

5

THE COW WITH
EAR TAG #1389

During the dairy market auction, I had heard several employees talking about the twice-weekly cull market auction. I was immediately curious to see how it differed from the dairy market sale we had just watched. Cull auctions sell animals who have been culled from the herd: those who have been deemed no longer productive or economically viable. Animals sold at cull auctions will typically either be transported directly to a slaughterhouse or be moved to a feedlot or farm where they are fattened first before going to slaughter.

A few weeks after the dairy market auction, I asked my close friend, colleague, and fellow geographer, Tish Lopez, whose work is on health citizenship and the militarization of aid in Haiti, to come with me to the cull auction. She agreed. What neither of us knew at the time was that this shared experience of attending the cull market auction was the first of many field excursions in what we have come to affectionately call *buddy system research*. In subsequent years, we've developed an intentional methodological practice of joining each other in our respective research activities—not as collaborators, in the sense of traditional collaborative research, but as "buddies," as companions accompanying each other into field and analysis research processes that are emotionally taxing, potentially physically unsafe, or otherwise troubling to encounter alone. We have written elsewhere about this practice as a way to push back against the individualizing neoliberal and masculinist logic of the university and "field research," advocating for acknowledgment and embodied practices that honor the deeply emotional and often traumatizing nature of research dedicated to uncovering injustice and structural violence.[1] When we attended the cull market auction, we were engaging in what felt intuitive for us as close friends and colleagues: I was anxious about going alone and I asked for companionship; Tish cared about me and my well-being and agreed to be my companion.

The cull market auction was held in the same auction yard as the dairy market sale. We set out midday to make it there in time for the 2:30 P.M. sale. We arrived a little after two o'clock and found that the auction had already begun—early, because there were so many "spent" cows to sell that day. There was no time for looking at the animals in the rear holding pens, though we did see an old gray horse waiting in a pen on our way inside. We went straight into the auction hall and sat in the front row near the auction ring's exit door.

As we entered the hall and found our seats, nearly every eye in the hall turned toward us. At the previous auction, the bleachers were fairly full with mostly men, but there were also some women and children. My dad—a seventy-year-old white-haired white man—and I blended in well enough. At this auction, Tish and I were the only women in the audience, joining just a handful of older white men who were referred to as the "meat buyers." Tish and I stood out substantially in this space: I wore jeans, a plaid button-down shirt (which a friend later told me was *hipster* plaid, *not* farmer plaid), canvas high-tops, and a billed cap to try to cover a particularly urban haircut. Tish also wore blue jeans and a cotton button-down shirt with black Doc Martens boots. We both hid our tattoos beneath our long sleeves, and Tish had taken out her facial piercings. As much as we tried to blend in, we were clearly out of place at this auction, and every time we turned to chat quietly to each other, I could feel a glance from men in the audience and, occasionally, from the auctioneer himself. No one talked to us.

I was so socially uncomfortable entering that space that it took me a few minutes to focus on the animals passing through the ring. As soon as I let my eyes rest on a cow in the auction ring, though, I was struck by the markedly different condition of the animals being sold at this auction in contrast to the dairy replacement sale. These were all severely worn-out cows—mostly black-and-white Holsteins—their bodies visibly destroyed by years of dairy production. Many of them, it turns out, were not more than five or six years old, though their bodies looked ancient. Their skin hung loosely on their hip bones and against their ribs. They were dirty, caked in mud, feces, and scabs. Many of them were emaciated and limping badly. Many of them had docked tails. Many of them had udders that were red and infected or dragging on the ground. Their eyes bulged, the whites showing. Mouths foamed with saliva. This was the look of the cull market auction: animals fearful, worn-out, close to death.

Loud bellows echoed through the auction hall. I was immediately overwhelmed, unable to focus on each individual animal because of the scale of the suffering, each devastated body blurring into the next.

Tish and I looked at each other. After years of friendship, we communicate a lot to each other in just a glance, a moment of our eyes meeting. We quickly looked away, knowing we just needed to focus on the animals in front of us, not thinking, not reacting, just watching. We were seated in the front row by the door through which the animals exited, so as each cow left the ring, we were able to look into her face. As I sat there and met the gaze of cow after cow, I felt deeply ashamed to be human. To be a member of a species that so systematically breeds, raises, uses up, sells, kills, and consumes not just cows but many other species felt sickening and unforgivable.

This feeling only intensified when a Holstein cow with ear tag #1389 limped through the door into the ring. She was small for her breed, and the impacts of her life as a commodity producer were easily legible on her body. Her tail was docked, her hide was covered in scrapes and abrasions, and she had an auction sticker with a barcode stuck to her side. Her frame was slight, and her ribs and hip bones protruded visibly beneath her skin. One of her back legs was not bearing weight (the source of her limp). Her udders hung to the ground and were red with mastitis. Compromised mobility and mastitis are common in cows used for dairy, especially those at the cull market auction, since both of these ailments frequently signal a cow's declining productivity.

Most of the cows at the auction that day were selling for $50 or $60 per hundred pounds of weight (by weight because they were all just one more stop away from becoming meat, their bodies disassembled and sold in quantity per pound). When the cow with ear tag #1389 entered the auction ring, the auctioneer started the bidding low—at $20 per hundred pounds. No one bid, and the price quickly dropped to $15, then $10, and finally to $5. No one bid. At seven hundred pounds, the cow with ear tag #1389 could not be sold for a mere $35, and the teenaged handler in the ring began to herd the cow toward the door. To me, she didn't immediately look much different from the other cows that had passed by, but the experienced meat buyers could see right away that she was not worth buying. Suddenly the audience erupted in a chorus of "uh-ohs," "oh boys," and "there she goes." Although I had my eyes glued to the cow in the ring, the seasoned meat buyers knew before I did what was happening. The cow collapsed, crumpling to the ground, in the ring. There was a momentary silence and then the auctioneer said, "Well, let her rest, I guess." They left the cow there in the ring and, not wanting to mar the efficiency of the auction, continued selling cows around her. Several cows were brought in one at a time, turned in circles, were sold and exited while the cow lay on the floor, her mouth foaming with saliva and her breathing labored.

While this was going on, my mind raced with frenzied thoughts: *Should I have bid? Was it too late to buy her? What would happen to her because she didn't sell? How would I transport her if I did buy her? Would she fit in my station wagon if I put the seats down? Where would I take her? Where would she live? How long could she live in our tiny backyard before our neighbors complained? What ethical questions were involved with financially contributing to the auction and were these outweighed by the good it might do to buy her and give her a different life? Why this cow and not the dozens of others I had watched pass through the ring?* I knew from my research already that many "downed" (nonambulatory) cows who may be close to death in the moment are often rehabilitated, needing only fluids, food, rest, and some basic veterinary care to perk up and recover. I wondered if this was the case with the cow with ear tag #1389. Thirty-five dollars was nothing to buy a whole cow's life; I had spent more than that on the tank of gasoline I had purchased to drive to the auction. But the practical details of buying her overwhelmed me as

I sat there, my body rigid, watching the scene unfold. The necessity of a transport trailer, the practicalities of quickly finding a large animal veterinarian and a sanctuary to take her in—these, paired with my lack of firsthand experience caring for cows, caused me to freeze in the moment, and I sat there and did nothing.

While I was going over and over these questions in my mind, a cow came into the arena who was spooked, running quickly and erratically around the ring. Her movements startled the cow with ear tag #1389 and she struggled to her feet, looking dazed. The two teenagers working the ring hurried to herd her out the exit door before she had the chance to collapse again. I caught her gaze as she left the ring, went through the door, stepped on the scale, and was gone.

Tish and I sat and watched another thirty or so animals pass through, sold on to slaughter, before the auction concluded. As we were leaving the auction yard, the animals who had been sold were already being loaded into extra-long transport trailers destined for the slaughterhouse. We got in the car, pulled out of the gravel parking lot and, as we turned out onto the country highway, we saw one of the auction workers standing up on a rung of the fencing that contained a group of cows. He held a thick metal rod and was yelling and striking one of the cows on the head and back.

As we drove home, Tish and I talked about the auction and cried. We both felt useless, experiencing keenly the limitations of our place as academic researchers in the world and as spectator-observers in the auction ring. Already, my research had a decidedly political arc to it, dedicated as it was to understanding the experience of what it meant to be a body living and dying for commodity production. There is no way that this work could not be politically and ethically charged in its approach and in what it produced. And indeed, as my fieldwork unfolded, my commitment solidified to write articles and a book that would be read both within and outside of the academy, with the hope of making an impact on the way people think about, and practice, our relationships with farmed animals. Within the academy, it often feels risky to foreground the political and ethical dimensions of animal life and to challenge some of the most basic assumptions about human-

animal relations that are foundational to the anthropocentric university (particularly at an institution like the University of Washington,
which brings in enormous financial support for the use of nonhuman
animals in biomedical research). This work—navigating the academic
waters, trying to make a case for research that challenges the uneven
power hierarchies between humans and other species—often feels
like an enormous task on its own.

When confronted with the urgency of each animal's life and fate
passing in an instant through the auction ring, this academic project
felt small and insignificant—especially when faced with the question
of whether to intervene and buy the cow with ear tag #1389 or to stop
the car and question the man beating the cow. How many cows would
pass through that ring on their way to the slaughterhouse, the dairy
farm, or somewhere else while I was completing this research or writing this book? How many animals' lives would this book do nothing
at all to change?

There are many animal rights activists who would have bought the
cow with ear tag #1389 or intervened on her behalf after the auction
was over. I've heard countless stories from advocates or activists who
followed their instinct to step in and then figured out what to do next.
At the auction that day and on the drive home, I envied this sponta-
neity, quick thinking, and determination. As an academic, I had been
trained as a careful and critical thinker and observer—dedicated to
the life of the mind, and this means that I do not make decisions easily,
instead ruminating on the possible outcomes, the ethical and political
implications, and the practical dimensions of a particular path forward
before acting. In this moment and for days, weeks, months afterward,
I rued this part of who I was (or had become).

This also raised questions for me about the politics of witnessing.
I have written elsewhere about the role of witnessing in academic research and the politics of emotion in response to witnessing animal
suffering; witnessing, as a political practice of documenting violence
and social injustice, has a long and important history as a mode of
political action in human and nonhuman contexts.[2] And certainly,
there is a need for research that witnesses violence and suffering, cre-

ating a tangible record in order to document, educate, and possibly prompt political transformation. But, witnessing can also be an ethically fraught and emotionally traumatic process precisely because it is dedicated to *witnessing*—watching, documenting—and not centered on doing something to change the conditions for those who are the subjects of violence. For the cow with ear tag #1389, my witnessing did nothing.

That night, after the auction, I had nightmares about the cow with ear tag #1389 as images of her played over and over again in my dreams: her body crumpling to the ground; lying there unable to rise; stumbling through the exit door; our eyes meeting as time froze for a moment before she went through the door and was gone. As soon as the auction opened the next morning, I called to ask about her. I explained that I'd been at the auction the day before, that I'd seen her not sell and collapse, and that I was wondering what had happened to her. Was she still there, available to buy?

"No," the man on the other end of the line said, matter-of-factly, "I know the one you're talking about. She was dead in her pen when we came in this morning."

Why the cow with ear tag #1389 occupied (and continues to occupy) so much of my mental energy and concern when *all* of the cows at the cull auction were on their way to die at the slaughterhouse I don't know. Maybe it was her suffering, so viscerally discernable in front of us in the auction ring. Maybe it was the fact that she had been so worn-out, so used up by the dairy production process that she couldn't even walk through the auction yard (the second-to-last stop before she would be slaughtered for meat). Or maybe it was because her collapse made her memorable and unique as she stood out from the regrettably forgettable stream of worn-out cows passing through the ring on their way to slaughter. Whatever the reason, the cow with ear tag #1389 haunts me. I can still, years later, see her face when I close my eyes. It is for her (and all the others like her) that this book is named and dedicated.

The paradox of the cow with ear tag #1389 is that, because she was in such bad shape, she was spared going through the slaughter process.

Because she didn't sell and because she died at the auction yard that night, she might have had a less violent (and perhaps, less frightening) death than the others who sold at the auction that day. Although, it's hard to say if dying of dehydration or disease is necessarily easier to endure than being slaughtered.

My visit to the cull market auction prompted two primary questions about the end of the lives of cows used for dairy: How are the animals who pass through the cull market auction slaughtered? And, what happens to the bodies of animals who die before they get to the slaughterhouse, like the cow with ear tag #1389?

Each of the worn-out cows who sold at the cull market auction continued on to the slaughterhouse—either stopping to be fattened up first or being transported directly to slaughter. It was completing a Master's thesis on slaughter practices in the United States that led me to the dairy project. And so, going into this research, I was familiar with the practices of slaughter and the laws governing the slaughter process. But the work I had done on slaughter hadn't faced the actual, embodied animals who were being killed. I had read industry guides, watched videos, and did other forms of textual research to understand the slaughter process. With this in mind, I'll now turn to slaughter, followed by rendering, to explain these key processes in the death of animals. In a typical trajectory of a cow's death, she would first be sent to slaughter and then what was left after the slaughter process (what wasn't usable for human food consumption) would be delivered to a rendering facility, where her remains would be transformed into other usable commodities.

SLAUGHTERING COWS RAISED FOR DAIRY

How are cows killed in slaughterhouses? As I was working on this project, I struggled with whether I needed to *see*, in person, animals being slaughtered to be able to adequately write about the process. I talked with many of my colleagues about it, as I was (unsurprisingly) having difficulty accessing spaces of slaughter. One colleague finally said to me, "Look, Katie. There's tons of work on slaughter already for you to draw on. And do you really need to see an animal be killed to talk about how and why it is an ethical problem? Furthermore, *should* you?" Practically speaking, she's right: there is a wealth of information in the form of text, video, and audio available documenting slaughter. In particular, there are already ethnographic works about slaughter-house labor by scholars and journalists to draw on, describing the process in detail. And there is also plenty of industry documentation about what should constitute "best practices" in slaughter. I found that looking at industry guidelines, federal and state laws, and government-documented violations of these laws illuminated quite a lot about slaughter as a routine industry practice.

As I considered the complexities of seeing animals be slaughtered, I was already struggling with the fuzzy ethical gray line between wit-nessing and voyeurism when watching animals suffering in the auc-tion yard. No matter one's political or ethical commitments in spaces of violence against animals, there is a certain level of complicity in this violence that occurs through standing by and watching animals die or be killed while doing nothing. In the academy, there is also a mascu-linist anthropocentrism embedded in the expectation that researchers should witness slaughter, or other forms of violence against animals, in their research. This links up with feminist critiques of masculinist con-ceptions of what "counts" as research and knowledge production. An-thropocentrism permeates animal studies scholarship that stands by and observes violence against animals; an implicit and underacknowl-edged belief that animals matter less than humans enables human re-searchers to think that the ends (writing about and sharing knowledge about violence against animals) justifies the means (observing while doing nothing to interrupt, prevent, or object to this violence). I count myself among this group in a way that has caused me to reflect on and question my own research practices. And yet, for researchers or jour-nalists trying to understand these practices and what impact they have on nonhuman animals, there is the question of how to access that kind of knowledge or information. Through my own difficulties accessing spaces of animal commodification, I came to understand that there are many ways of building knowledge about dairy production that do not involve such overt complicity in harming animals. I decided I would proceed without witnessing slaughter in person, although later, inadvertently, I encountered the slaughter of three animals in an entirely different context.

But first, a description of the way the majority of animals are slaughtered in the United States today: as animals are loaded up to leave the farm or the auction yard, destined for the slaughterhouse, the transport itself is a major source of stress for these animals who are often conveyed hundreds of miles to slaughter with infrequent food and water.[3] In some parts of the world, live export of animals is common and animals will travel for days and sometimes weeks as live cargo on ships (the Australian wool industry has come under scrutiny for its live export practices involving sheep who are transported on ships to the Middle East where they are slaughtered for meat).[4] Farmed animals in the United States are generally transported by truck and sometimes train, enclosures that expose them to extreme heat and cold, and many animals die in transport from heat exhaustion or by freezing to death. Transport trailers are frequently packed tightly and the stress of confinement and the conditions of the trailers can cause many animals physical, psychological, and emotional distress.[5]

Transport trailer crashes involving trucks carrying animals on interstate highways throughout the United States and Canada are not uncommon. In November 2016, a transport trailer crashed in Lubbock, Texas, injuring the driver and injuring or killing several dozen cows.[6] In June 2015, there were news reports of a crash of a transport trailer in Ohio carrying twenty-two hundred piglets from South Carolina to a farm in Indiana. Records showed that three to four hundred piglets died in the crash, with many more injured at the scene.[7] Accidents like these populate the news across the United States and Canada, often recounted as part of traffic reports that warn of long delays and road closures for interstate travelers. These accidents are obviously a welfare issue for animals, as are the conditions under which animals are shipped to slaughter.

Slaughter in the United States is predominantly performed in large, industrialized facilities. The increasing consolidation in agriculture more generally is echoed in the slaughter sector, with smaller facilities closing in favor of more industrial and mechanized systems of killing. The slaughter of chickens, for instance, is almost entirely automated, with chickens shackled upside down on conveyor belts, where their throats are cut by a mechanized blade before they are dunked in a boiling water bath to remove their feathers.[8] Facilities slaughtering cows are also increasingly mechanized, industrial processes reliant on a (dis)assembly line model, with each mode of killing and butchering compartmentalized into single, repetitive labor tasks.[9] It is believed that Henry Ford actually developed his assembly line model for automobile manufacturing from observing the efficiency of the disassembly of animal bodies in the historic Chicago stockyards.

Cows arrive at the slaughterhouse and are herded through chutes that lead into the killing floor. Temple Grandin, who has a PhD in animal science, works with the meat industry to redesign slaughterhouses with the aim of improving the experience of animals moving through these spaces. In Grandin's description of her work, she discusses how she has used her own experience as a person with autism to understand how animals experience the world.[10] From her distinctive perspective, she redesigns slaughterhouses (namely, the parts of slaughterhouses where the animals are still alive) to reduce the level of fear animals experience leading up to their deaths. She recognized through observing cows, for instance, that they balk at the sight of certain objects, lights, or movements, and so she designed curved chutes that prevent the animals from seeing what is ahead of them.

Many laud Grandin's work for improving the welfare of animals in slaughterhouses and for her concern about the well-being of animals leading up to their deaths. Indeed, her work has not only improved the experiences of animals but reducing their fear responses in the lead up to slaughter also helps the industry to operate efficiently, allowing more animals to be killed with fewer slowdowns. Grandin also seems to operate as a figure who assuages consumer uneasiness about killing animals for food. Throughout my Master's thesis research on slaughter, and even while completing my research on dairy for this book, it became commonplace for friends, family, and acquaintances to bring up

Grandin's slaughterhouse designs when they heard about the subject of my work. The frequency with which Grandin was mentioned was likely, in part, a result of an HBO movie called *Temple Grandin*, starring Claire Danes, which was released in 2010 and turned into a pop culture reference that helped Grandin become a household name. What's interesting about Grandin's work is that, although her slaughterhouse redesigns mitigate some of cows' negative experiences leading up to slaughter (what they see, how the chutes are designed), animals in slaughterhouses are still butchered and they still smell the blood of those killed ahead of them and hear the bellows of those moving through the chutes.

When the cows reach the killing floor, they enter the "knocking box," where they are restrained while the worker whose job title is the knocker renders the animal unconscious using a captive bolt stunner (the current tool of choice). This tool is generally powered by an air compressor and drives a steel bolt into the animal's forehead. The animals are then moved along the line where another worker shackles one of their hind legs and hoists them upside down on a conveyor belt where their throats are cut and they're bled. Then they're moved along as their bodies are efficiently disassembled, one step at a time. In the best cases—and in Grandin's recommendation—animals are rendered unconscious on the first blow by the knocker, but sometimes it takes several tries, and it is not uncommon for animals to proceed to shackling, hoisting, and throat cutting while they are still conscious. Because of the speed of the line, this means that animals are sometimes skinned alive or have body parts removed while still conscious, as revealed by undercover videos available to view online.

Questions about visibility and access to information related to slaughterhouse practices are a recurring topic for scholars and activists dedicated to improving the plight of farmed animals. Slaughterhouses are typically located in rural areas, and the structures in which slaughter is performed are routinely warehouses without windows, distant and difficult-to-access spaces that conceal the practices of slaughter from the general public. Timothy Pachirat, in *Every Twelve Seconds*, his book about slaughterhouse labor, explores how this invisibility works to obscure the violent impacts of the process on both laborers and animals in slaughterhouses. He articulates, for instance, how consumers can forget the violence of meat eating because they do not, themselves, have to perform the act of killing. And, after working in a Nebraska slaughterhouse for five months as part of his dissertation research, he explains that workers, too, are alienated from the act of killing, caught up as they are in the very real ways in which they are also exploited subjects of violence.

Pachirat describes a *politics of sight*, exploring how institutionalized violence, like that operating in the slaughterhouse, is enabled by what and how we see—by what is concealed and revealed and how. One of the jobs Pachirat performs in the slaughterhouse is that of a quality-control worker, a role that allows him movement throughout many different parts of the slaughterhouse and affords him near-total viewing access to the inner workings of the plant. Through this role, paired with his other jobs in the chutes where live animals are driven into the knocking box and as a liver hanger in the coolers, he articulates the complexity of the politics of sight. It isn't just a matter of making certain practices visible; there is a more multifaceted politics of concealment and surveillance at work.[11] He acknowledges that the act of making processes like slaughter visible *can* bring about social and political transformation (and this is, in fact, part of his aim with the book); however, this is not always the case. He writes: "Even when intended as a tactic of social and political transformation, the act of making the hidden visible may be equally likely to generate other, more effective ways of confining it. We have already seen, with the slaughterhouse quality-control worker, how isolation and sequestration are possible even under conditions of total visibility."[12]

The geography of the slaughterhouse, in ways similar to the auction yard, is spatially designed to promote efficiency and to move a constant stream of commodified living, dying, and dead animals through these spaces in ways that discourage thoughtful contemplation on the significance of this commodification and killing. One of the most useful parts of Pachirat's text is, in fact, a set of diagrams of the slaughterhouse that visualizes the compartmentalization of labor happening

in the slaughterhouse that makes it possible for a liver hanger to forget that the livers he is hanging each belonged to an individual, living, breathing being not long before reaching him. In part, it is this
compartmentalization operating at the slaughterhouse—through the
mechanization and industrialization of food production—that trans-
lates into how consumers can remain willfully ignorant of, and dis-
tanced from, the farm, the auction yard, the slaughterhouse, and the
rendering plant. But, importantly, as Pachirat points out, even in cases
where there is total or near-total visibility of the process, the spatial design, the speed of the line, and the general culture of commodification
take the sight before you and make it abstract.

Related to this abstracting effect, and resonating with my own experience of the auction yard, Pachirat's work highlights the monotony
of slaughterhouse labor—the speed of the line, the repetitive nature of
disassembling a once-living being, and the profound boredom that
comes with this work. There's something important and noteworthy
about this boredom and the way it functions—for the worker in the
slaughterhouse or for the observer at the auction yard. Boredom and
monotony have a way of glossing over the reality of what is before you;
they dull the mind and the senses in particular kinds of ways. Being
bored in the auction yard made my mind wander; it made me do things
like count the animals or record their conditions without really seeing them; it made me impatient for the auction to be over; and, most
importantly, it made it much more difficult to feel the weight of each
sale and what it meant for each individual cow passing through the
ring. In the slaughterhouse, Pachirat describes the level of monotony
as not only abstracting from the cow but also enacting its own kind of
violence on the worker—in the effects on the body of repetitive stress
injuries and injuries caused by rushing to keep up with the speed of the
line and, also, in the effects on the mind of doing such mind-numbing
labor.

Before she was enrolled in college or graduate school, Yolanda Valencia, a friend and colleague of mine, worked in a slaughterhouse when
she first moved to Washington from Mexico. She took the job because
it was good pay compared to the other options she had available to her,

and, unlike seasonal field labor (e.g., orchard labor) in eastern Washington, slaughterhouse work is year-round, and thus a stable source of work and income. She described to me that, as a petite woman, more than anything else, she struggled with the speed of the line and physicality of the work. She used her whole body to move the weight of the meat and make the cut for which she was responsible before it moved along the line to the next worker. For months when she began working there, she would have to soak her hands before work just so that she could open them and move her fingers, so excruciating was the impact of the repetitive motions required for the job. When she was on the line, she told me, she wasn't thinking about where that piece of meat had come from; instead, she was thinking about making her cut and keeping up with the line without cutting herself, fearful of getting fired or having her coworkers have to help her complete her job.

Slaughterhouse laborers are overwhelmingly people with low earning potential, people of color, formerly incarcerated people, and undocumented immigrants—all enormously vulnerable populations who are relegated to doing some of the most difficult, dangerous, and dirty work in society. The violence against both workers and animals in slaughterhouses is why structural critiques of the systems that produce this violence are so important. Capitalist economies require the commodification of the animal and extraction of human labor in order to grow and perpetuate the accumulation of capital. Focusing on the violence only against the animals or only against the workers obscures the ways in which structural conditions produce violent effects for *all* involved. In many ways, it is so much easier to blame the individual workers or a few bad corporations, instead of taking a hard look at the ways in which capitalism is implicated in deep-rooted forms of exploitation of humans, animals, and ecological systems. It is also easier to believe that we can make a more just world by simply buying the right things—this is why there is such a growing market for so-called humanely produced animal-derived foods. Even consumption-driven forms of veganism (the kind of veganism that is focused solely on a consumerist lifestyle and not the kind of veganism that is a politics and ethics rooted in a fundamental rejection of anthropocentrism and speciesism) are premised on this idea that we can "buy our way to more ethical living."

Some leaders in the animal advocacy movement fall into this trap—that we should put our energies into "voting with our dollar"—without acknowledging the foundational problems with the system itself that produces, demands, and recirculates those dollars. Capitalism is a fundamentally extractive system reliant on social relations that extract labor and energy from the humans and animals on whom the system relies, as well as their very bodies. The further up the economic ladder individuals climb, the less exploited they are by the extractive nature of the system and the more possible it is to imagine that they got to where they were by some feat of individualized, personal excellence. The American Dream is premised on this upward climb—the idea that anyone, with enough hard work, ingenuity, and pluck, can succeed at becoming middle or upper-middle class (in other words, that—in theory—anyone can become active or overactive consumers to continuously stimulate the economy by buying nicer cars, bigger houses, more expensive gadgets, and so on). But capitalism requires a large and active labor force (the proletariat, according to Marx) to fuel the accumulation of capital, and this global labor force is highly gendered and racialized, with women and people of color occupying the least well-paying and most exploitative of society's jobs. Slaughterhouse labor is no exception in this regard, although more people who identify as men tend to work in slaughterhouses (74.3 percent of all slaughterhouse laborers in the United States).[13]

Common responses to the exploitative dimensions of how free market capitalism operates often promote smaller-scale, alternative modes of food production. Small farms and boutique slaughter operations are actively promoted as the ethical alternative to industrial modes of production. For this reason, I was particularly interested in exploring how these places operate. Consolidation in the food industry, which often eliminates local options for slaughter, causes small farmers who raise animals for meat to transport animals over increasingly long distances to their deaths. Most animals raised on small farms, then, are still subjected to the stress of transport and the reality of the industrial slaughterhouse. More and more often, though, large slaughter-houses have minimum numbers they will accept for slaughter from a single farmer, and so farmers raising only a small number of animals do not meet these minimum requirements and must find an alternative method of slaughter.

An additional obstacle for small farmers in the slaughter process is that farmed animals sold for meat (excluding birds) must be slaughtered in a USDA-approved facility. USDA approval involves costly inputs, including machinery and inspectors and it is often prohibitively expensive for small-scale farms to implement onsite. To respond to this problem, the mobile slaughter unit emerged as an affordable, small-scale method of slaughter.[14] Mobile slaughter units are trucks that have had their trailers converted to a tiny slaughter facility. They travel to the farm, slaughter small numbers of animals, and can be approved by the USDA as an acceptable method of slaughter, which is necessary to sell meat in US markets. These mobile facilities are growing in popularity in the Pacific Northwest (and beyond) and are celebrated as a way to improve the welfare of animals (since they do not have to leave the farm) and as a way to return control and autonomy to the farmer in an industrializing agricultural landscape. Thus, in theory, mobile slaughter units could offer a somewhat improved method of killing animals for food: eliminating transport can be a not-insignificant improvement, and the smaller scale of the slaughter operation could translate to more care being taken in the slaughter process. Certainly, if animals slaughtered by these units do not have to be transported to reach them, that is an important improvement. But in the actual practice of slaughter, my research has not convinced me that mobile slaughter units offer an alternative that is radically different from other forms of slaughter.

On a cloudy Pacific Northwest afternoon, my partner, Eric, and I were shoveling woodchips at Pigs Peace Sanctuary in Stanwood, WA, where we regularly volunteer. We were in the TLC area of the sanctuary, where pigs live who need extra care and who might be easily injured if they were living in the main herd. While we worked, pigs ambled around the TLC area, and some curious about what we were doing stopped by for a back scratch. The TLC area is on the south edge of the sanctuary grounds, directly adjacent to a tiny, single-lane road. Across the road sits a small family farm that raises animals for meat. Over the years of visiting Pigs Peace, I've seen various species of farmed animals pastured there: sometimes they raise steers for beef and other times they are raising pigs for pork and bacon. In years past, their long shedlike barn had been filled with chickens. On this particular afternoon, Eric and I watched as a large white truck pulled up on the narrow road, parked, and opened its trailer door. We continued to work while the driver and passenger talked and laughed with the two farmers. We didn't know that this was a mobile slaughter unit.

We stopped working to watch as three pigs were led out, squealing, to the truck. One of the men retrieved a firearm from the back of the truck and, on the ground behind the truck, aimed and shot one of the pigs in the head. The shot did not kill the pig or render him unconscious. In front of the other two pigs, the wounded pig thrashed around on the ground letting out piercing shrieks. The other two pigs panicked and tried to escape but were contained roughly by the farmers. The man fired again. The pig's shrieks only intensified, and the pigs in the TLC area where we were working squealed and ran to the far edge of their enclosure, as far from the dying pig as they could get. The man quickly fired a third time and the pig went still. He shot the other two pigs quickly, killing them each on the first shot. The pigs were then, one by one, hoisted up on the back of the truck where they were bled, eviscerated, and dismembered.

We watched, stunned at what we had just witnessed. That first pig had suffered—the pain of two gunshot wounds—before dying, and the other two pigs suffered the terror of watching one of their own killed, painfully, in front of them before they, too, were killed—one and then the other. In what is considered the best mode of slaughter, we witnessed the botched killing of one—one-third of the animals there. These were professional butchers. They were not rushing. This should have been an example of slaughter done flawlessly.

Watching the slaughter by this mobile unit, I tried to imagine if it would have been radically different if all three pigs had been shot correctly on the first try. Surely it would have reduced the amount of prolonged pain the first pig felt and possibly lessened the trauma for the other two pigs of seeing one of their own suffer before them. But this experience prompted more fundamental questions about farming and killing animals for food. That is, I realized that the conversation that we might have more often was maybe less about *how* humans in the United States kill animals for food and more about *whether* they should raise them and kill them for food at all.

This is a big question that challenges some of the most fundamental beliefs we have about farmed animal species. Raising animals for food on farms relies on a more than ten thousand–year history of domestication. Familiar narratives about animal domestication are that it is often celebrated as one of the most significant feats of human progress and evolution and that is was a boon to animal populations as well; in this view, domestication offered protection for the newly domesticated species in an otherwise life-threatening landscape.[15] The argument advanced by folks like Jared Diamond in *Guns, Germs, and Steel* that animal domestication was mutually beneficial in historic times has such powerful appeal that it is reproduced in the present to create a commonsense narrative about continued animal domestication: that were animals not bred and raised for food, they would cease to exist at all; thus, farming them ensures the survival of their species.

David Nibert, in his book *Animal Oppression and Human Violence: Domesecration, Capitalism and Global Conflict*, offers a counter reading of the history of domestication, arguing that domestication was the precursor to the foundation of a capitalist economy that fundamentally violates and dominates both nonhuman and human others. The process of domestication and the subsequent millennia of selectively breeding animals for traits that *humans* deem desirable; the extraction of biological material in the form of milk, meat, and eggs; the containment of animals in enclosures whose quality and size are determined entirely by humans—these processes, as features of domestication, shape and appropriate fundamental dimensions of animal life. Slaughter, then, represents the last mechanism of appropriating animal life. When the cow has ceased to be productive in generating milk or calves as commodities, the last of her life energies are redirected into by moving her to slaughter, whereby she is transformed, through killing, into yet another commodity form as meat. Her life, which was never her own, is claimed once more in this moment of killing. But although she is now dead, this is not the end of her body's commodification.

RENDERING ANIMAL REMAINS

After the slaughterhouse, animal remains not sellable for human food consumption are transported to a rendering facility. Animals, like the cow with ear tag #1389 and others who die before they reach the slaughterhouse (and thus cannot enter the human food supply), are also typically routed directly to a rendering facility. These facilities break down the bodies of animals into usable components to be sold and transformed into new commodity products. These products include fertilizer, pet food and farmed animal feed, biofuel, soaps and detergents, lubricants, pharmaceuticals, toothpaste, paint, and many more everyday products.

After the cull market auction where I encountered the cow with ear tag #1389, I wanted to know more about rendering to understand where the remains of her body ended up. I discovered that there was a rendering plant not far from where I lived in Seattle. I had driven by it hundreds of times over the years without knowing what the structure contained. I called the company and spoke over the phone to a friendly and helpful man—Kevin at Urban Rendering.[16] I explained my project and that I was interested in learning more about the process of rendering the bodies of dead farmed animals. Kevin explained that they were not involved in what he called full-body rendering; since this was an urban rendering plant, he told me, they historically did only partial remains rendering. Now they handled mostly recycled cooking oils from restaurants. He wanted to help, though, so he gave me the number of a man he knew who worked as a deadstock hauler, driving his truck around to farms to pick up dead farmed animals and deliver them to rendering facilities, for a fee.

Dave, the deadstock hauler, was gruff when he first answered the phone: "Yeah, I'm Dave."[17]

"Hi Dave, my name's Katie Gillespie. Kevin at Urban Rendering gave me your number and suggested I get in touch. I'm a researcher at the UW, and I'm studying the dairy industry. I'm interested in the rendering process—you know, what happens when an animal dies on a farm. Kevin thought you might be willing to chat."

"Oh yeah, Kevin, he's a good guy." Dave's voice softened. "Interested in rendering, huh? Don't get much of that, in my line of work. Well, let me see what I can tell you," he said, his voice trailing off. Then he continued, as he gathered his thoughts. "Well, I mostly haul dead horses. When a large animal like a horse dies on the farm, it's hard for most folks to dispose of it. For horses and cows, sometimes pigs, you need a front loader. Burying a large animal like that isn't practical, so rendering's usually the best and easiest option. Haulers like me get paid to pick up the deadstock and drop it at the rendering plant."

"Do you ever haul cows?" I asked.

"Not usually. Now and then there's a cow. Mostly I get calls for horses. But they're real similar. You have to load 'em, drive 'em, and drop 'em."

"And are you familiar with what happens at the rendering plant?"

"Yeah, I mean, I don't usually hang around." He chuckled and then paused before laughing again, "I have to admit it's awfully odd for a lady to be interested in rendering! Rendering is nasty business. I've seen grown men walk into a plant and lose their lunch. The smell is . . ." He paused to think of the word: "Powerful."

And, indeed, this was a sentiment I heard echoed across conversations with people who had worked in, or were familiar with, rendering. During my ethics review process, for instance, one of the reviewers of my application mentioned in a face-to-face meeting that she had worked for several months at a rendering plant in Washington.

"It was summer," she said, "and the smell was so bad, I'd get to work every morning and throw up. Eventually, after being there for a while, you'd adjust to the smell every day, but it was always a shock to my system when I arrived in the morning. I didn't last long. Just a few months before I left."

After talking to Kevin and Dave, I continued contacting rendering facilities, looking for one that would allow me to visit and see the process of how animal remains are rendered into materials for new products. Rendering is a transnational industry with large, international corporations overseeing localized rendering plants. Darling International, for instance, is an international corporation that owns numerous brands and operates across North and South America, Europe, Asia, and Australia. Among their North American locations is a rendering facility in Tacoma, Washington (just south of Seattle).[18] The rendering facility where Kevin works, it turned out, was one of the smaller companies in the area. But there are large global companies (like Darling International) and national or regional companies as well that make up the rendering landscape in the Pacific Northwest.

I talked with a general manager, Kurt, at one regional corporate renderer.[19] Although Kurt could not, ultimately, gain approval from headquarters to allow me to visit the facility, he did his best to help provide me with information to elaborate on the rendering process, and he shared resources produced by the industry explaining how rendering is done, as well as general statistics about rendering. He also answered more specific questions I had. I had heard in casual conversation that rendering sometimes deals with not only animals who die on farms or in slaughterhouses but also animals killed on roads and even animals (like dogs and cats) euthanized in shelters. I asked Kurt if they covered these types of situations.

Kurt replied, "We do work with the state or cities and accept certain animals that are killed on roads. It's limited to only deer, elk, and farm animals. We do not accept animals from shelters. I don't know if any go to other rendering companies. Most that I know of are incinerated through pet cemeteries."

"And do any of the rendered materials go into food products for human consumption or only animal feed/pet food or nonedible items?" I asked, having heard conflicting ideas about what exactly rendered materials could be used for.

"Rendered products are inedible and not sold for human consumption. Rendered products *are* used in animal feed." Kurt then also referred me to some online sources, such as the National Renderers

Association, to learn more about not only where rendered materials end up but also how rendering is done.

To fully explore rendering is a much bigger project than I could cover in the scope of the research for this book, but I wanted to at least understand the general operations of rendering as a peripheral industry related to dairy production.[20] I thought about the cow with ear tag #1389 and wanted to know how her body had been disposed of and how it was transformed and commodified beyond the point of death.

What I found through reading the materials Kurt sent me and chatting with a few men in the industry was that although folks like Dave, the deadstock hauler, describe rendering as a fairly grisly industry, it is also one of those lesser-known industries that society simply could not do without. Rendering transforms the remains of farmed animals who die before slaughter, the remains left over *after* the slaughter process, and in some places, the remains of animals killed on roads (passively termed *roadkill*, although there are other methods of disposing of these animals that are employed in some states — like composting, for instance). Without rendering, the landscapes in which humans live would be quickly overrun by the enormous amount of waste from farming animals and from other animal deaths. According to the National Renderers Association, rendering plants process fifty-six billion tons of waste in the United States and Canada each year, so much, in fact, that "if all rendered product was sent to the landfill, all available space would be used in four years."[21]

Approximately *half* of the slaughtered body of the cow — bones, blood, fat, some of the internal organs — is considered inedible to humans and goes to rendering after the slaughterhouse.[22] Animals who die on the farm or at the auction yard make up a small but not insignificant percentage — 4.5 percent — of rendered bodily remains; the overwhelming majority of rendered material is what is left over at the end of the slaughter process.[23]

The process of rendering begins with the "raw material," which includes the whole bodies of dead animals, or dismembered remains, fat, and bones. When this animal tissue arrives, it is first ground up to break it down into bits that are appropriately sized for the cooking process. Most rendering plants in the United States utilize what is called a continuous-flow cooking process that enables large quantities of by-products from the meat, egg, and dairy industries to be processed and separated into discrete new commodities. This steam-cooking process generally takes between forty and ninety minutes, melting the fat to separate it from the bone and protein, and ultimately removes most of the moisture from the tissue.[24] Next, the fat is further separated out via a screw-press device and moved into storage for transport and sale and the eventual use in goods like soap and other body-care products. The remaining material is composed of some excess fat, minerals, and protein, which are all further processed through grinding and separation into animal feed, meat, and bone meal, blood meal, and other products that are stored for transport.[25]

In addition to dealing with bodily remains and waste after slaughter, rendering also serves a purpose in eliminating potentially harmful bacteria that are often found in the bodily remains of animals prior to the rendering process, including salmonella, listeria, and campylobacter. These microorganisms are killed in the high-heat rendering process. One major concern in processing animal remains into animal feed—especially in the context of cows—is the spread of bovine spongiform encephalopathy, commonly known as mad cow disease. The disease is spread through feeding the remains of infected cows to other cows, a practice that may seem bizarre, but the drive for economic efficiency and low production costs has made it commonplace to feed farmed animals the rendered remains of other farmed animals. Rendering, as it currently stands, does not guarantee killing bovine spongiform encephalopathy, which is why the Food and Drug Administration prohibits feeding even small amounts of by-products of ruminants (like cows) to other ruminants.[26] Bovine spongiform encephalopathy and other diseases that affect farmed animal populations and also pass from farmed animals to humans (like avian influenza, for instance) are increasingly urgent issues as animal agriculture intensifies, demanding the growing confinement of animals in close quarters where diseases spread rampantly. The potential for disease transmission extends through the death of the animal and beyond into the disposal of their bodies, which makes the method of this bodily disposal an important health issue (not to mention an ethical one).

Because of the staggering numbers of animals who are brought into being, who labor, and who are slaughtered in the food system, rendering is an essential industry for managing the bodily remains of so many once-living beings. Rendering, as a largely invisible industry itself, is a key mechanism in concealing the death of animals for food and the disposal of billions of them globally each year. Thus, while rendering on its own serves the crucial function of managing and utilizing bodily remains, it is also entangled with industries—like dairy—that raise, commodify, and kill animals for food.

These systems of slaughter and rendering—killing and disposal—say something unsettling about how farmed animals are valued in the United States: not as singular beings with lives of their own whose deaths are grievable, but as bodies whose worth is attached to their ability to be grown, used and commodified, killed for food, and then rendered to eke out the last bit of economic value from the commodity afterlives of their bodies. This is the predominant conceptualization of farmed animals, animals like the cow with ear tag #1389 at the auction yard. The cull market auction highlights this logic. Indeed, sitting in the cull market auction, I had never felt this realization as keenly, perhaps because the space was so starkly dedicated to efficient economic exchange that would facilitate the final phase of farming animals: the killing and disposing of their bodies. Perhaps it was the stripped-down efficiency of the sale, the buyers who bought dozens of animals at once based on the promise of their remains converting seamlessly to meat, the sea of gaunt and worn-out bodies—already like ghosts of themselves—passing through the auction ring. It was in this state—reflecting on death and killing, disposal and waste—that I set out toward my next set of research sites: the farmed animal sanctuary.

6

SEEKING SANCTUARY

After a month of visiting auctions and the farm—spaces that were emotionally difficult to inhabit—I felt relieved to be driving south, alone on Interstate 5 to California, to visit two animal sanctuaries: Farm Sanctuary and Animal Place. These sanctuaries describe themselves as places of rehabilitation, animal care, and education about alternative ways of living with other species. As I drove, I felt lighter as I put mile after mile between me and the auctions, the farm, the mobile slaughter unit killings. I looked forward to a week of visiting the sanctuaries, learning about the animals who lived there, and talking with the people who formed these spaces.

It was June, and the Central Valley of California was already sweltering in summery heat. Farm Sanctuary is located off the main roads past a series of almond orchards and dairy farms. Just before reaching the sanctuary, I passed Wackerman Dairy, a former dairy farm that rents the land out for grazing animals from other farms. On another side of the sanctuary's property line was a ranch where a large herd of steers were pastured and, as Ryan, one of the sanctuary's animal care staff, would tell me during my visit, sometimes the animals would jump the fence into the sanctuary and sometimes Suzy (one of the cows in the main herd) would jump the fence into the rancher's pasture.[1] Cows are notoriously difficult to keep contained, as they are able—with a little incentive—to leap six foot fences or figure out other creative ways through fencing meant to contain them.

"What happens to the animals who jump the fence into the sanctuary? Do they get to stay at the sanctuary?" I asked naively.

"Well, sadly, we have to return them," Ryan replied. "We have to maintain good relationships with our neighbors, and we want the farmers around us to return the sanctuary animals to the sanctuary whenever they are the ones to jump the fence."

What a difference a fence can make, I thought. Such a visually simple device of enclosure and property delineation: a tangle of wire and wood cutting across the landscape, marking out territory and the differential ownership of animal bodies on either side. On one side, the animals bred or bought to produce milk and meat, their daily care dedicated to ensuring their ability to produce beef and ultimately a livelihood for the farmers who own them. On the other side of the fence, a radically different understanding of animal life: these animals had been routed out of the commodity circuit, each with their own story of how they came to the sanctuary, their care now intended to foster lives lived with as little harm to them as possible.

Aesthetically, though, the two different sides of the fence looked indistinguishable: they both looked like what one might picture when imagining the family farm with rolling hills of pasture, farm buildings clustered in one area, and animals roaming the land. In fact, Sue Donaldson and Will Kymlicka worry that the aesthetic similarity between many sanctuaries and the idyllic vision of the family farm subtly reinforces that the farm is where farmed animal species belong and that modes of care common in farming settings are how these animals thrive.[2] The casual visitor or observer may not notice or fully grasp the radically different conceptualizations of farmed animal species that occur on either side of the fence. And yet, at the same time, there is what anthropologist Elan Abrell terms a meaningful "symbolic power" in creating the material conditions where this different understanding of animals plays out.[3]

Farm Sanctuary is a nonprofit organization dedicated to animal protection and rehabilitation, education and outreach, and legislative and corporate reform. With sanctuaries in Watkins Glen, New York, Orland and Acton, California, and most recently, a New Jersey location, it is the largest farmed animal sanctuary organization in the United States. Founded in 1986 by Gene Baur and Lorri Houston, the first animal they took in was Hilda, a sheep they found lying on a "dead pile" behind Lancaster Stockyard in Pennsylvania where they were documenting inhumane conditions for animals. They were examining the pile of dead animals when Hilda lifted her head and looked at them. Baur and Houston quickly rushed her to a veterinarian where she recuperated, needing only water, food, and some basic care to make a full recovery. Hilda lived for eleven years at Farm Sanctuary's New York shelter before she died there in her old age.

When Baur and Houston first took Hilda in, they tracked down the person who had dropped her at the stockyard. Confident that this was a gross animal welfare violation, and prepared with evidence from their presence at the stockyard, Baur and Houston attempted to press charges. However, they were quickly informed that these actions constituted "normal animal agricultural practices" and, under Pennsylvania's common farming exemption, these kinds of actions were exempt from the state's anti-cruelty laws. This led the newly formed Farm Sanctuary to perform further investigations and lead a campaign against Lancaster Stockyards. After organizing demonstrations and getting Lancaster Stockyards to adopt a "no downer" rule (which meant that animals would be euthanized if they were too weak or sick to be sold), time passed and, gradually, the company went back to their former lax animal welfare practices. Farm Sanctuary incorporated as a Society for the Prevention of Cruelty to Animals in the state of Pennsylvania in order to have authority to enforce animal welfare laws given that other humane and law enforcement agencies were unwilling to take on the stockyard. After a humane investigator from Farm Sanctuary documented a nonambulatory cow at the stockyard, Farm Sanctuary was able to file charges for cruelty. This time, the sanctuary won the case and the stockyard was the first in the United States to be convicted of cruelty to animals.

Of course, as mentioned earlier in the book, farmed animal welfare legislation remains—decades later—inefficient in coverage and insufficiently enforced, and so, one mission of sanctuaries like Farm Sanctuary and Animal Place is legal reform. These sanctuaries work for the passage of state and federal legislation to improve protection for farmed animals. At the same time, they engage in education and out-reach about veganism, informing the public about the plight of farmed animals, introducing them to the animal "ambassadors" at the sanc-tuaries, and providing informational materials to encourage visitors to go (and stay) vegan. Some animal advocacy activists and theorists suggest that the "vegan conversion" strategy of the animal liberation movement is not a particularly effective method for making large-scale change: not only does it individualize the problem and solution, but also, many people who go vegetarian or vegan go back to eating an-imals later on, and it does not address the broader structural issues at work.[4] And yet, in talking with many people over the years who have visited sanctuaries, sanctuaries *can* inspire people to dramatically change their consumption practices and to think differently and more expansively about the kinds of relationships of care that are possible with farmed animals.

Farmed animal sanctuaries are commonly dedicated to the three-fold mission of refuge (which they often frame as rescue), education, and advocacy. As a result, contrary to the way I was received (or rather, not received) by most farms, the sanctuaries were happy to have me visit and ask any questions I had. Larger sanctuaries, like Farm Sanc-tuary and Animal Place, have internship programs where interns can live at the sanctuary for an extended period of time (usually more than one or two months) and volunteer in animal care, education, or out-reach/advocacy. These internships are also an important part of the mission of sanctuaries as they help to train a new generation of animal advocates in sanctuary work—both for the expansion of sanctuary efforts and for the longevity and sustainability of the sanctuary itself. Sanctuaries need long-term commitment, and dedicated human care-givers are necessary to carry on the work of the sanctuary once the original founders have died or are too old or ill to do the daily work of the sanctuary. Smaller sanctuaries with less formal infrastructure and few or no staff members are particularly vulnerable in the event that the primary person or people involved in sanctuary work are injured, become ill, or die.

Before embarking on research about sanctuaries, I believed that their primary mission was the rehabilitation and care of animals. Certainly, there are some sanctuaries where this is the case—sanctuaries where education and advocacy are either not part of their mission or where these are a less central focus. At Farm Sanctuary and Animal Place, however, taking animals in was only part of the equation. Staff from each sanctuary acknowledged the centrality of the animals' care and the importance of creating a safe space for the animals who live there. But they also acknowledged that widespread change would likely not occur through only providing refuge.

Ryan at Farm Sanctuary pointed out that "there are simply too many animals to ever believe we could rescue them all from the food industry. Rescuing individuals doesn't make a drop in the bucket in terms of the larger problem. But it makes a difference for the individual animals we take in and they become ambassadors for their species— for all the others who don't make it to sanctuary."

Anne, an intern at Farm Sanctuary, also emphasized the broader project of advocacy when she told me, "Honestly, it would be irresponsible if we dedicated *all* of our time and energy to animal care when there is so much work to be done on the bigger picture. *Of course* we take excellent care of the animals here and we work hard to ensure that they have the best life we can give them, with as little suffering as possible. But we also recognize that this is not enough if we want to make any real change in the structures that cause this suffering to begin with."[5]

I was struck by their simultaneous concern for the animals in their care and their awareness of the structural dimensions, and capitalist economic logics, which maintain the commodification of farmed animals. This pragmatism is rooted in an understanding of the scale of animal agriculture, awareness of the limited resources (space, money, etc.) of sanctuaries, and knowledge of the way the experience of singular animals is deeply connected to structural forces of capitalism, tradition, and species hierarchy.

The first animal I met at Farm Sanctuary was Norman, a tiny calf— not more than a few months old—quarantined in a small enclosure in the shade beneath a grove of trees where staff and visitors passed

FIGURE 6.1 Norman, Farm Sanctuary (Orland, CA)

regularly to visit with him. Norman was seized by Livingston Animal
Control from a man who was raising him for slaughter. He was the subject of neglect: just one month old, he was underweight and suffering
from pneumonia when he was taken in by the sanctuary. Norman was
born in the dairy industry and, as a male, was considered a by-product
of dairy production. In an effort to eke out some profit from this "by-
product," calves like Norman are raised for four to six months for veal
or eighteen to twenty-four months for beef. When I met him, Norman
was bottle-feeding and recovering well from the pneumonia but was
being kept in isolation as a precaution to ensure he made a full recovery. I followed up later and learned that after he recovered he went on
to a permanent home with a member of the Farm Animal Adoption
Network in southern California. When they can, sanctuaries like Farm
Sanctuary and Animal Place will find adoptive homes for those animals they feel can successfully thrive in a new home.

Ryan showed me around the area where the pigs were lying in the
barn in the hay, large industrial-sized fans blowing on them to keep
them cool. Next, we entered the goat area. The goats were curious,
and several of them came up to us when we entered their barn. After

that, we met the turkeys and ducks and chickens. I asked Ryan what they did with the eggs the chickens laid. He explained that they hard-boiled the eggs, crushed them into a mash (shells and all), and fed them back to the chickens. Egg laying, particularly laying an egg every single day, puts strain on a hen's body and leaches essential nutrients from the bird. Feeding them the eggs, he explained, was the easiest way for the hens to replenish the nutrients that are lost in laying an egg. Hens raised for eggs will lay an egg almost every day of the year. This accelerated laying schedule is the result of generations of breeding chickens for egg production, selecting the more prolific layers, breeding them, selecting the most prolific of their offspring, breeding them, and so on. I once thought it was just a natural trait of this particular species to lay so many eggs. But when I took the time to think and read about it, it seemed odd. What bird (or other animal) ovulates three hundred or more times a year? Most birds lay a few eggs in a single clutch once—maybe twice—a year. Humans typically ovulate once a month. Thus, an animal, like the chicken, who ovulates once a day is constantly taxing her body.

This acceleration of bodily processes is the case in the dairy industry as well. Cows have also experienced a drastic increase in the volume of milk they produce as a result of breeding. For instance, between 2007 and 2016, the number of pounds of milk produced by each cow in the United States has increased by 13 percent.[6] This increase is even more striking over a longer period: in 1930, the average cow in the United States produced approximately 12 pounds of milk per day, but by 2016, milk production per cow has increased to approximately 62 pounds per day.[7]

Farm Sanctuary, as well as Animal Place, house animals according to species. Even within species, they will further segregate animals: bovine animals at both sanctuaries, for instance, were separated into a "main herd" and a "geriatric herd" to keep older animals from being injured by younger ones. Or, at Animal Place, when I visited, the pigs were divided into "big pig" (pig breeds raised for food) and "potbellied pig" areas—posited again as a way to keep smaller, more vulnerable animals safe. Judy Woods, at Pigs Peace, in contrast, allows all species of pigs (and a few other animals, like sheep and a llama) to coexist in the main herd. On one of my first visits to Pigs Peace, I remember asking, "Don't the big pigs ever hurt the little ones?"

"Size has nothing to do with dominance," Judy explained. "You can have the littlest pig be dominant over the biggest pig—bossing him around. They make their own hierarchies and their own friends. Who am I to tell them that they can only be friends with, or live with, this pig or that pig just because they're the same size?" Instead, the divisions and separate areas at Pigs Peace are organized around the level of care needed. For instance, Judy keeps an area for medical quarantine and another called the TLC area, for pigs who need ground that is more level than the main field, for instance, which is nearer Judy's house so she can keep a closer eye on their needs. These divisions are imperfect, and Judy sometimes has to move or separate pigs when conflicts arise among them. She also takes into consideration the relationships formed between pigs in the main herd if a pig must be removed from there for medical care, often moving not just the pig in need of medical care but also that pig's closest buddy.

Donaldson and Kymlicka point out that species segregation remains a feature of most sanctuaries.[8] Although there are reasons to do this that are oriented around ideas about good animal care, Donaldson and Kymlicka explain that, "on closer inspection, segregation may be based more on human assumptions (or convenience) than on responsiveness to the needs and desires of individual animals."[9] They point out that, in sanctuaries with "intermingled populations" (where different species are allowed to live in the same enclosures), like VINE Sanctuary in Vermont, cross-species relationships emerge and the risks associated with housing different species or sizes of animals together are ameliorated by providing "sufficient space for shelter and hiding spots."[10] Shortly before the publication of this book, I had the opportunity to travel to Vermont and visit VINE Sanctuary, a place I had been reading and hearing about for years for its unique approach to creating sanctuary for formerly farmed animals. Nestled in a hilly forested landscape, many of the nonhuman residents at VINE comingle in heavily treed areas. The human residents at VINE have designed different forms of housing and shelter to accommodate the needs and preferences of the residents in these shared, multispecies spaces. Because the landscape itself is varied (much of the land is forested, with some areas more open and others grassy or rocky), the nonhuman residents exercise a fair amount of choice in how they live and with whom. For instance, some of the chickens and roosters prefer to roost in the trees at night instead of seeking shelter in their coops—a practice that prompted sanctuary cofounders pattrice jones and Miriam Jones to reflect on their role in offering the conditions for meaningful and livable lives for residents with different individual needs and desires.

Sanctuaries, then, pose a possibility for exploring other nonnormative ways of creating livable spaces for formerly farmed animals that do not reproduce farming models of species segregation, farm-based practices of care, and highly uneven power relationships between human caretakers and animal residents. Of course, a certain level of imbalance in power is always going to occur in a sanctuary space where animals are still captive—although the effects of captivity can be mitigated in various ways.

One way to mitigate captivity and transform knowledge about the care of farmed animal species is to incorporate animals in the decision-making process. pattrice jones described an instance when a crucial decision was being made: "We stood in the barn surrounded by sanctuary residents, as we like to do when making important decisions. (Miriam and I have always believed that decisions about animals ought to be made, insofar as possible, in consultation with animals. If that's not possible, the next best thing is to be in physical proximity to animals like those you're thinking about, so that you don't make the mistake of treating them as abstractions.)"[11] Even in the sanctuary setting it is possible to view the singular animal as an abstraction and to make decisions or judgments about care based on generalized species norms; however, sanctuaries have the potential to be sites where care is radically reimagined by taking the time to acquire knowledge about singular animals and their needs and also about their unique relationships with others living at the sanctuary. Abrell writes, "Beyond their symbolic value for inspiring struggle toward a better future, sanc-

FIGURE 6.2 The main herd, Farm Sanctuary (Orland, CA)

tuaries perform the essential task of working through the difficulties and contradictions of manifesting that future—the pragmatic labor that must be done in order to achieve more radical transformations in human-animal relations."[12]

Back at Farm Sanctuary, after showing me the pigs, goats, chickens, turkeys, and ducks, Ryan led me up behind the buildings, past where they care for any sick animals; contrary to the way Pigs Peace is organized with the medical quarantine front and center and visible to all visitors, Farm Sanctuary positions the sick animal area behind the scenes, away from visitors' view. These contrasting decisions are interesting both in terms of animal care and in terms of how the sanctuary is curated to facilitate a particular visitor experience. In both cases, there are good reasons for positioning infirm animals where they do. At Pigs Peace, Judy positions them closest to the house so that she passes them many times a day and can check on them. As a former emergency room nurse, Judy has spatially ordered Pigs Peace to mimic the logic of triage: animals with the most emergent care needs are positioned most proximately, while those who are not urgently infirm but who may just need more routine daily care are positioned a bit farther out, and the pigs who are most independent in their need for human care roam in the main herd. This logic is motivated by the pigs and their unique needs for care; since Pigs Peace is open only during summer months for visiting, visitors and their experience are a lower priority.

At Farm Sanctuary, the positioning of the infirm animals "behind the scenes" reflects a desire to keep them and those of their human caretakers actively providing care in the moment from being disturbed by visitors. But keeping infirm animals out of view of visitors also curates a particular kind of visiting experience, one that presents a thriving community of healthy animals, thereby reinforcing the idea of the sanctuary as refuge.

Ryan and I walked up into the far pastures where the main herd of cows and steers spend their days. They grazed the green grass that grew around the edges of a large pond and nibbled on hay that was delivered daily to feed them during the summer months when the pastures were brown and dry. Several of the cows were friendly and curious and walked right up to us as soon as we entered the pasture. I gasped as they came close to us. They were huge! This was the first time I had been face-to-face with cows without a fence separating us and they towered above me. There is something about cows behind a fence that somehow mediates their height.

"Watch your feet," Ryan warned. "A thousand pounds stepping on your toes can crush your foot." I looked down at my canvas sneakers and shifted my feet away from the giant hoof just inches from my foot. As I was looking down, one of the cows swung her head to swish flies off her face, and I could feel her head graze my shoulder. "Watch out for swinging heads, too!" Ryan took a step back. I followed suit and then, from an arm's length, one after another I scratched the cows' necks and behind their ears. Ryan told me the stories of many of the animals, including those around us—Oliver, El Niño, and Harrison, as well as Phoebus, Blake, and Sixer, a trio of young Jersey steers who were new to the sanctuary and were brought in along with two others who were taken in by Animal Place. Most of the steers in the main herd were not from the dairy industry but came from various farms in California and beyond where they experienced conditions of routine industry use or, what Animal Control would formally define as, neglect and abuse. Several of the animals kept their distance and, if we moved closer, they would back up, so we let them be.

The lasting emotional trauma some animals experience from their pasts is visible at every sanctuary I have visited. Although animals are safe and are treated with care and gentleness at places like Farm Sanctuary, Animal Place, and Pigs Peace, it is, for many animals, a long process of healing and learning to trust humans. Some animals remain wary of humans for their whole lives, preferring the company of others of their species (or species other than humans) or a select few favorite human caregivers. Sadie, whose story began this book, was an example of a cow who, even after more than a decade at Animal Place, remained wary and fearful of many humans—although she had close relationships of love and care with her caretakers, such as Marji Beach.

I had arrived at Farm Sanctuary around midday, and by the time we were done with the initial tour, I got the key to the cabin on top of the hill, where I would spend the night. The cabin was a large single room, which included a bed, a sitting area, a small kitchen and eating area, and a separate bathroom. I took off late afternoon and headed into town, where I stopped for some groceries and dinner. When I got back to the sanctuary in the evening, the temperatures were slightly cooler, and I wandered around at dusk, observing the animals as the sun began to set, the cows swishing their tails to swat at flies, and lizards racing across the path in front of me.

The People Barn, an education center and gift shop, was just next to the cabin, and I wandered around outside to read the informational boards that communicated to visitors the violence of farming and, in particular, the ways in which animals are raised in increasingly industrialized spaces. There were models of standard confinement systems for visitors to look at, touch, and imagine inhabiting: a veal crate, a gestation crate for sows, and a battery cage for hens. It was after hours for the People Barn, so no one else was around. And as I stood there alone, I felt a wave of grief, and my skin got goosebumps thinking about what it would be like to live my life in one of those confinement devices—unable to stretch my arms, or turn around, or interact with others. And this is the purpose of displaying these devices: to evoke an embodied visceral response wherein visitors meet the animals and learn about their personalities and then imagine those same animals (those named, individual, singular animals) confined in these devices. For instance, faced with the tangible veal crate as an artifact, it was difficult not to imagine Norman, the Holstein calf, confined in that crate, and to imagine the impact that extreme confinement would have on his life, well-being, and psyche.

There are many ways that spending time at sanctuaries for human visitors or volunteers can be a profoundly healing, hopeful experience. The first time I visited Pigs Peace (my first sanctuary experience), I had a powerful sense of *coming home*. Seeing animals living with a fair amount of freedom to express their species-specific behaviors, in community with others of their species, spending their days (to a certain extent) as they choose feels momentous and important. But sanctuaries can also be emotionally taxing places to visit because they are real, tangible reminders of what nearly all farmed animals don't experience, confined as they are by industrial modes of farming. Indeed, each animal at the sanctuary represents millions (and billions) of others who will never live in this way. As I walked back to the cabin, I thought about this tension.

That night, I climbed into bed—exhausted—and lay there in the dark, sweating. The heat was nearly unbearable, and I decided to get a wet washcloth for my head. I turned on the light, and as I put my feet down on the floor, I saw three enormous brown spiders skitter across the floor in front of me. I carefully avoided the spiders and walked to the sink in the kitchen to wet a cloth and, there, on the wall by the fridge, was a large black widow spider with the characteristic red hourglass shape on its back. I grabbed the cloth and my cell phone and ran back to the bed, leaping into the middle and pulling the blankets up so they weren't touching the floor. I called my partner, Eric, on the phone and, with some panic, explained my predicament.

Like many people, I'm somewhat afraid of spiders. I also believe that spiders and other insects should be able to live their lives without humans killing them out of fear or disgust. My usual routine at home when I find spiders in the house has always been to carefully relocate them outdoors with a glass and a sheet of paper to cover the glass. But being at the sanctuary—a safe haven for many different kinds of creatures—made me pause. It somehow felt wrong to relocate these spiders whose home was this cabin, and I their visitor for just a couple of nights. Being at the sanctuary reframed how I thought about space and whose space this was. In other words, relocating the spiders would infringe on them living their life, uninterrupted, in their home. And so, I lay in the middle of the bed with the blankets piled up around me in a kind of wall, talking to Eric on the phone until I fell asleep.

I haven't always been afraid of spiders; this was a learned behavior. When I was a tiny kid, I used to lie in bed at night and watch the spiders crawl across the ceiling of my bedroom. I had no fear of those spiders. It was only later that I developed a fear of them. I couldn't tell you where or how exactly I learned to be afraid of spiders. It's possible it was partly the summer vacations I spent in rural Virginia with my great-aunts who lived in a farmhouse on some land in the middle of horse country. The house was like a fortress—closed up tightly, air-conditioned in the heat of summer, and reeking like cigarettes (one of them smoked inside, with the windows closed)—and excursions outside involved severely delivered warnings about the creatures that might get us: the ants, the ticks, the spiders (black widows, no less!), the snakes, and so on. I grew up believing that the outdoors at the Virginia farm were a hostile environment—to be avoided at all costs.

Sometimes my mom would convince my sister and me to come out and do some gardening with her, and I remember, now and then, she would find a black widow during her gardening and call us over: "Girls, come and look at this! Isn't she beautiful?" And I would be torn, looking at this creature and thinking she *was* beautiful and, at the same time, hearing the terrifying warnings from inside the house echo in my head.

Or, sometimes the great-aunts would decide they needed to kill all the weeds in the gravel driveway, so they would get out there with a backpack sprayer and soak the weeds with Roundup. Or they would zap a colony of ants on the back patio with a spray can of poison. The garage was a testament to their fear of nonhuman life. One cabinet, in particular, contained an arsenal of weapons against the creatures who might possibly dare to infringe on what they viewed as *their* little patch of land: weed and insect killers, mouse and rat traps, rat poison, fly traps, mosquito repellant.

As we sat at the kitchen table with the great-aunts inside their fortress, they would tell us stories of the animals against whom they had waged war during their life at the farm—the groundhogs and rabbits who ate their garden, the deer who were a constant nuisance (though we only ever saw them grazing out in the grass by the tree line), the foxes they claimed they used to shoot at with a rifle. They even reminisced about a little dog they'd had at one point whose feces so disgusted them that they would wash his rear end off with Lysol every time he came back in the house.

In retrospect, it was not the environment there that was hostile to us but, rather, the great-aunts who were hostile to the environment. Still, their words and warnings terrified my sister and me, and so we spent most of our time in the cool, smoke-filled house, drinking southern sweet tea and playing the card game Crazy Eights on the spotless white living room carpet.

It might, then, have been those summers in Virginia that helped me develop a fear of spiders—or it might have just been growing up in a world generally hostile to most insects and other creatures deemed "pests." Whatever it was, I felt keenly at the sanctuary that I wanted to change how I felt about—and thus, acted toward—the often misunderstood and despised creatures of the world: insects, rats, pigeons, and so on. This started with *not* relocating the sanctuary spiders. And later, at home, this extended to changing how I interacted with spiders and other critters I encountered; it extended to viewing what I had previously thought of as *my* home as a space where my partner and I cohabitated with many nonhuman others.

The way some of us may feel disgust or fear about spiders or rats is important for understanding how categorization of species permeates the way many humans organize and shape the lives and deaths of others. The species hierarchy and categorization that allow the killing of a rat or a spider as a pest also enable the systematic use of the cow for dairy. But these categorizations (pest, food, pet, research subject) rely on a more fundamental category—the category of "the animal"—which relegates nonhuman bodies (and often, certain human bodies, too) to a lesser, subordinate status that is justification for exploitation. Although humans themselves are animals, the figure of the animal—the *category* of the animal—is a site of violence and subjugation. The sanctuary can prompt questions about what it means to transform these dominant categorizations and to appreciate the embodied and emotional lifeworlds of species regularly relegated to being *food*. Shifting our relationships with nonhuman animals requires a fundamental change in how humans share space with and conceptualize other species.

But even sanctuaries cannot guarantee animals live lives entirely free from suffering. Many of the animals who come to sanctuaries can continue to suffer as a result of their previous treatment and use in the industry and as a result of their breeding. Sadie, for instance, had a chronic limp from her broken leg and hip and she suffered emotionally from the trauma and fear she experienced in her encounters with humans prior to arriving at the sanctuary, although she learned to trust both certain humans and other cows in her long life at the sanctuary. But the earlier parts of her history lingered even after she had been living at the sanctuary for nearly a decade.

Issues arising from breeding practices in animal agriculture cause some of the most common ailments among animals living in sanctuaries. At Farm Sanctuary, for instance, sanctuary workers often explained the difficulty of caring for ducks and turkeys whose genetic makeup (through selective breeding) made it so that they were unable to stand, so weighed down were they by the rapid growth of their bodies for meat production. The consequences of breeding programs in animal agriculture include animals whose bodies cannot sustain their rapid growth, who are predisposed to mobility issues, and so on, and sanctuaries, in contrast to farms, see animals *age* and have to deal with the day-to-day realities of the aging farmed animal body. Because of these realities, much of the work of sanctuaries is to try to minimize

animals' suffering as much as possible. Consequently, they will sometimes make the difficult decision to euthanize an animal whose suffering cannot be mitigated.

Julie at Farm Sanctuary, commenting on the ethical dimensions of
having the responsibility for making these kinds of decisions about
how animals live and die at the sanctuary, said to me:

> You know, I wonder sometimes, is it even kind for us to keep
> animals alive in these bodies that are really uncomfortable for
> them? Is that okay? I don't know. The stories that they tell do so
> much good in the world. You can meet Valentino [a steer] here,
> and he's got interests and opinions. When you can see that these
> are individuals with interests and worth on their own. And then
> I also think that they are happy to be alive and living lives of their
> own here, you know? And we can manage their pain in all kinds
> of ways. We are doing a good job keeping these animals as com-
> fortable as possible in the bodies that they've got. I hope they are
> happy to be alive, but I can't talk to them. There's no way to know
> that what we're doing here is 100 percent what they would want.[13]

These decisions and ways of viewing animals involve complex disability politics and questions about communication across species.
Sunaura Taylor documents how animals with disabilities are conceptualized within animal agriculture in her book *Beasts of Burden*: she
discusses the way many farmed animal species are born with physical
mobility and growth issues as a result of their breeding for commodity production; she also documents injuries and disabilities caused to
farmed animals as a result of how they are used in spaces of production (the cow with mastitis or mobility issues as a result of intensive
milking, for instance). Within this context, Taylor also explores how
the industry itself (as well as consumers) treat animals with disabilities in the food industry as problems to be managed or disposed of
(usually through killing) so as not to contaminate the so-called normal animals routed into slaughter for food.[14] Sanctuary caregivers understand the challenges in mobility and longevity caused by farmed
animal breeding (and also negligent treatment or abuse on farms), but

there is another layer of complexity to the disability politics at work here. Eli Clare, in his book *Brilliant Imperfection*, pushes back against normative ideas about embodiment, care, and difference, arguing that our very ideas of *normalcy* and *naturalness* ought to be eroded. Clare is troubled by the often unquestioned imperative to *cure* people of their disabilities; in a nuanced way, and without arguing wholly against notions of cure, he writes, "I think about myself and all the disabled people around me—acquaintances, friends, coworkers, neighbors, family members, lovers, activists, cultural workers. I think about what we offer the world—comedy, poetry, performance art, passionate activism, sexy films, important thinking, good conversation, fun. I think about who we are and the ways in which our particular body-minds have shaped us. Who would we be without disability?"[15] What Taylor's and Clare's work prompts us to ask is at once how animals' lives and relationships may be shaped by disability—and how often these disabilities may be caused by breeding and common industry practices (certainly these practices should be sites of critical and ethical concern in the ways they cause harm)—and also how we might reframe thinking about disability not as something to be unquestioningly eradicated but as different forms of embodiment that may shape, in beautiful and unexpected ways, who an individual is and how they move through the world.

And the best methods of care may be more difficult to discern in contexts where it isn't easy for animal caregivers to communicate in clear ways with the individuals in their care, as Julie articulated above. But sanctuary caregivers also point out that there are many ways that animals communicate that do not involve speech, and animals often communicate quite clearly their feelings about how they are being treated and what is going on around them. Thinking more creatively and expansively about forms of connection, embodiment, and communication may offer even deeper forms of care to flourish in sanctuary settings and transform the relationships that exist in these spaces.

With these questions of care, disability politics, and connection in mind, there are other forms of human power and bodily control over animals at sanctuaries that pose complex ethical questions. Animals in sanctuaries are still captive; their "rescues" involve humans deciding that the sanctuary is better for them than any alternative. Many animals in sanctuaries are sterilized, as most farmed animal sanctuaries maintain that reproduction should not be allowed there, in order to maximize resources for other animals coming out of industry spaces.[16] Most animals at sanctuaries are also subjected to various kinds of veterinary care and treatment that they may not necessarily choose, although this care is provided with the animal's well-being in mind. And again, when their well-being is compromised by declining health, injury, or age, human caretakers will often make the decision to euthanize animals at the sanctuary. So many decisions are made for animals in sanctuary spaces, and this power differential between human caretakers and animal residents is an ethically fraught issue.

Cofounder of VINE Sanctuary Miriam Jones expresses ambivalent feelings about being in a position of care for formerly farmed animals:

> We use the term "as free as possible" deliberately, as fences, enforced routines, involuntary medical procedures and regimes (including everything from forced sterilization to force feeding), and other impositions certainly do not comprise a free state of being for those on the receiving end. Those of us in the sanctuary movement routinely make decisions about the animals in our care (and under our control) that we, as ethical individuals, should find extremely problematic. How can one justify taking the reproductive ability away from another individual? How about taking the possibly-fertilized eggs of a broody hen, or penning up an active cow in a stall because she needs rest for an injury? The answer is simple: we justify these decisions because the alternatives are unacceptable. We live in a world that requires the rescue of members of certain species because other members of our own species will hurt and kill them if we don't; we do what we need to do, as ethically as possible, within the context of that reality. We also know—those of us who work with formerly farmed animals—that for most of them, survival on their own is an impossible goal.[17]

I thought about these ethical complexities as I left Farm Sanctuary in Orland and made the two-hour drive south to Grass Valley, where Animal Place is located. The landscape around Grass Valley is very different from Orland. As I neared Grass Valley, the temperatures cooled slightly from the more densely forested areas I was driving through, and Animal Place is itself part forest and part expansive green fields.

When I pulled into the driveway, there were several people working in the front garden and they showed me into an old farmhouse that had been converted to office spaces for the sanctuary staff. There is where I met Marji Beach, the education director at Animal Place, and we prepared to head out to the fields to meet the cows and see the sanctuary. Marji wore tall galoshes and, taking one look at my flimsy canvas high-tops, disappeared for a moment, returning with what she explained were rattlesnake guards. They were pieces of tough fabric with something hard embedded in them that wrapped around my calves, extending down onto the top of my foot, and fastened with buckles. These, she explained, would protect against any surprise rattlesnake bites. "Rattlesnakes, huh?" I asked, trying to sound nonchalant. But I'm pretty sure my eyes widened like round saucers.

"Yeah, there are rattlesnakes here. We don't see them *that* often, but better to be safe!" Marji said cheerfully.

"Yes! Better to be safe," I agreed as I buckled the rattlesnake guards around my shins and reminded myself about yesterday's commitment to being more inclusive and caring about my interactions with other species (even those who terrified me).

We set out down the driveway into the sanctuary, passing by an expansive vegetable garden. Marji pointed it out as we walked: "That's our newest project there—it's a vegan micro-farm." Marji explained that vegan farming, also known as veganic farming, does not utilize any animal products: no manure, no bone meal or fish meal fertilizers or animal-based soil additives.[18] Instead, they use all vegetable matter to enrich the soil. This method of farming is low impact, entirely plant based, and sustainable.[19]

"Don't you have a whole lot of manure, though, that you could use here on the farm?" I asked, "I mean, I can see why you wouldn't want to *buy* any animal-derived soil additives or enrichers and support systems of animal exploitation, but what about the poop that the animals here are producing anyway?"

"We're not ethically opposed to using the manure here because the animals don't really care," Marji brushed a fly away from her face, "But because we want to be a model for other small farms and individuals, we choose not to use the manure. To show that it's possible to have successful food production without the use of animals. So, yeah, you're totally welcome to go through the farm and pick some chard, kale, peas, strawberries—all sorts of good stuff—on your way out."

"Thanks. This is cool," I replied, "You know, I actually teach a class on animals, ethics, and food, and the students frequently bring up questions about how you can have a successful, sustainable farm without using animals. So this would be such a useful model to teach about."

"Yeah, that's what we're hoping. Sometimes people don't get why we don't use manure. It's so distant. Like, you get it with meat because obviously the meat comes from a dead animal and when you talk about dairy and eggs, people get that—that eventually the animals are all killed and there's all these negative side effects. But it's really three steps removed when it comes to manure, and people don't realize where it really comes from—like the pig farms where animals are confined and horribly treated. So it's a whole new education process for us as well."

Animal Place was founded in 1989 by Kim Sturla and Ned Buyukmihci, beginning with a sixty-acre parcel of land in Vacaville, California. There was no house (or other structures) on the land and, living in a trailer at first, they built the sanctuary completely from scratch. Sturla and Buyukmihci had started the sanctuary because they were frustrated by the disconnect they saw in how people related to farmed animals. Buyukmihci was working as a veterinary ophthalmologist and professor at the University of California, Davis, and Sturla worked at a local dog and cat shelter, and they both wanted to create a space where people could connect with, and learn about, farmed animals in a setting not dedicated to their use. Shortly after they moved to the new property, the dog and cat shelter where Sturla worked took in a farm pig who they named Zelda. The shelter couldn't find a home for Zelda where she wouldn't be slaughtered and eaten, so Sturla and Buyuk-mihci took her in as Animal Place's first resident.

Over the years, the sanctuary grew and gained support, eventually acquiring the six hundred–acre space in Grass Valley. Over time, Grass Valley became the location where around three hundred permanent animal residents live and where education and visiting is focused. The Vacaville space has become Rescue Ranch, the location that handles new rescues and facilitates adoptions by people who have the commitment to taking in a formerly farmed animal.

As Marji showed me around the sanctuary, I met many of the animals—the rabbits, the chickens, turkeys, and geese, the pigs, the goats and sheep, and, finally, the cows. As we were walking along one of the paths, a large wild jackrabbit crossed our path. I had never seen a jackrabbit before, and it was mesmerizing to watch how the large hare hopped effortlessly, his powerful back legs propelling him forward. He stopped to look at us and his ears oscillated like giant satellite dishes, his nose and whiskers twitching, before he hopped off and disappeared into the brush.

As we crossed the field to join the geriatric herd where Sadie lived with Elsa and Howie, Marji told me Sadie's story—the story that opened this book. Sadie, Elsa, and Howie lay close to each other in the tall grass, clearly choosing to hang out together despite the expansive pasture they had to explore. As Marji told me Sadie's and Elsa's stories—both had been used for dairy, but in very different circumstances—she alternated between them, scratching both behind the ears and in those hard-to-reach places. They were clearly very fond of Marji and nuzzled her hand and against her chest as she scratched and patted them.

Elsa was a small, creamy brown Jersey cow with a gentle disposition who lived for fifteen years at a Waldorf school farm, where she was artificially inseminated, gave birth to calves, and was milked regularly. Marji did not name the specific Waldorf school that Elsa came from, but Waldorf is an educational program originating in Stuttgart, Germany, in 1919 and is dedicated to fostering intellectual, practical, and creative skills in children. Some Waldorf schools have farms as part of the educational experience—children learn practical skills related to growing food and raising animals for meat, dairy, eggs, and fiber. As students age, their roles and responsibilities on the farm increase, and they are involved with more animal care and farm labor. For instance, at Sunfield Farm in Port Hadlock, Washington, second and third graders are responsible for tending their own flocks of chicks, while eighth and ninth graders are involved in raising larger animals, like cows.

In contrast to traditional farms where the calves would be taken away at birth, at Waldorf, Elsa's calves stayed with her for a time to nurse before they were sold. Jersey cows are popular among Waldorf educational farm programs because they are small, easy to take care of, docile, and the children love them. When Elsa was fifteen, she was no longer getting pregnant and, after the school's purchase of four draft horses, they decided that the cost of feeding and caring for Elsa if she was no longer reproductively viable was too much. They were planning to sell her at auction when someone at the school contacted Animal Place and asked them to take her in. Normally, the sanctuary does not take in animals who people are getting rid of out of convenience or because they don't want them anymore; the sanctuary tries to prioritize what, by the law, would be considered the most urgent and overt cases of abuse, neglect, and cruelty from factory farm scenarios. Admittedly, understandings of cruelty, abuse, and neglect vary widely in these kinds of contexts, and these kinds of decisions will often be determined by sanctuary staff on a case-by-case basis. Certainly, abandoning an animal like Elsa out of convenience can be understood as a form of neglect; neglect is a state of being without care, and a former caregiver ceasing care for an animal and sending them to slaughter would lead to a severing of that caring relationship. Although the school itself would not see this decision as neglect since it follows a normalized tradition of disposing of an animal when they are no longer of use to humans, the sanctuary staff saw her life and a responsibility for her care in very different terms.[20] They agreed to give her a home.

Her first night at the sanctuary, Elsa bellowed all through the night. The next day, when she was introduced to the herd, sanctuary staff discovered that she had what Marji called a "sassy" personality—she was bossy and rude with the other cows and steers, head-butting them and scaring giant Howie so that he had to hide behind Sadie. Sadie stood up to her, though, and by the time I met them a couple of years later in the geriatric pasture, they appeared to have worked out their differences and enjoyed having each other nearby as they lay there, munching on grass.

As we stood in the pasture talking, Marji said, "It's just heartbreaking to think that this animal who had trusted these people for fifteen years, who was loved and cherished, who had a name, she still has that name—Elsa—was going to go to the slaughterhouse. There's all kinds of betrayal that goes on in the farming industry, but there's something very provocative about *this* type of betrayal—really betraying an animal you've claimed to have loved for fifteen years."

And it *was* heartbreaking to think about Elsa being sold for slaughter because she was no longer able to reproduce. It was also troubling to think of her being artificially inseminated repeatedly to keep producing calves so that students could learn how to milk a cow. And what was this act of raising a cow, using her till she was "spent," and sending her off to slaughter teaching children exactly? What was it teaching children about the reproductive body—that certain bodies have value only as reproductive entities? What was it teaching children about the aging body—that in old age we send our loved ones off to die somewhere else (at the slaughterhouse, in the nursing home, etc.), where we don't have to deal with the inconvenient realities of aging or death? And what was it teaching children about the value of animals—that animals (even those we love and cherish) have value only insofar as they are useful and convenient to us but that, in the end, they are disposable? What did it teach children that one person at the Waldorf school intervened and routed Elsa to sanctuary instead of the cull market auction—did it create a rupture in the framing of the cow as disposable?

Spending the day with Marji at Animal Place prompted many questions for me about the role of education. I was moved, thinking about the sanctuary as an educational space that teaches adults and children alike a different kind of ethic about our relationship with farmed animals — namely, that farmed animals are beings with rich inner lives of their own, that they should not be subjected to lives dictated by prioritizing human interests of commodity production and consumption, and that, given a place of their own, they can and do flourish in community with other animals and humans committed to their care. In a society so dedicated to teaching us that animals are here to be eaten, worn, experimented on, and kept as pets and entertainers, the sanctuary embodies an alternative conceptualization of how animals fit into multispecies social worlds.

On the long drive home, I thought about the work of places like Animal Place, Farm Sanctuary, Pigs Peace, and VINE Sanctuary, and I thought about the dominant modes of education about animals that they were working against. This would have to be my next site of research — to try to understand how, in the United States, we learn to think of animal use as normal.

7

DOUBLETHINKING DAIRY

While I was doing the research for this book I was invited by a colleague to accompany her to her child's social justice–focused elementary school in Seattle where she was going to talk to her child's class about the reason their family did not go to zoos. The class was taking a school trip to the zoo that day and, beforehand, she explained to the kids the ethical issues with zoos and captivity more generally. We then went along as chaperones for the zoo trip. It was my first visit to a zoo in probably fifteen years, and I was curious about what Seattle's Woodland Park Zoo was like and how it compared to other zoos I had visited in the past.

Because this was a school trip, we had a zoo employee educator guide us through parts of the zoo. The part that stood out was when we entered the farmed animal area, and we shuffled into a small building where the educator showed us what raw wool fiber looked like and how a sheep's fleece was transformed into felt or wool for yarn. He passed around some pieces of fleece and encouraged the children to take a sniff of the fleece. Then he held his nose and said, "Peee Ewww! Sheep really smell bad, don't they, kids?" And without missing a beat the children gagged and held their noses. And just like that, the group of children learned to be disgusted by the natural aroma of another species. How different their educational experience might have been in that small moment if the educator had said, "Make sure to smell the fleece—doesn't that smell interesting, so earthy and complex?"

We continued on from the lesson about sheep and were given a tour around the farmed animal exhibits. In front of each exhibit the educator asked what the kids knew about the farmed animals. In front of the cow pen, he asked, "What do we use cows for?"

In unison, the children shouted, "Milk!"

"That's right! Milk!" the educator replied, with no mention, of course, of beef or the reality of dairy production.

In front of the sheep pen, he asked, "And what do we use sheep for?"

In unison the children shouted, "Wool!"

"That's right! Wool!" the educator grinned.

In front of the chicken area, the educator asked, "What about chickens—what are they used for?"

"Eggs!" the children replied, immediately.

The educator agreed. Finally, in front of the pigpen, I thought, "This is it. How can he avoid mentioning meat in the case of pigs?"

He seemed unconcerned, though. Without hesitating, he asked, "What do we use pigs for?" The children were silent. Not one of them knew. The educator paused and then said, "Kitchen scraps! Pigs are terrific for eating unwanted kitchen scraps."

And then we moved on, leaving the farmed animals behind, their education about farming complete for the day. Ultimately, the omission of a conversation about meat was somewhat irrelevant. Even if the zoo educator had informed the children that pigs were turned into bacon, hot dogs, and pork chops, the outcome would likely have been the same. Without the details of slaughter, and what the act of slaughter means intrinsically about the subordination and appropriation of other species for human interests, the children would have likely just accepted this fact and moved on without taking in its significance.

Most adults have become expert at glossing over inconvenient or unsavory truths. In fact, it's at least partially necessary for our continued ability to function. If we didn't engage in acts of forgetting or ignoring at least *some* painful truths about our impact on the world—the scale of suffering and destruction we are implicated in just by living as active consumers—many of us might not be able to function, experiencing depression, anxiety, and existential crises about the meaning of life and suffering. And so, we ignore, we gloss over, we actively forget.

The painful truth about the violence involved in farming animals for food, then, is something against which many people shield themselves and their children. This level of proactive ignorance and forgetting is one way we learn to relate to other species—an active not relating and a form of magical thinking that if we don't think about it (the violence), maybe it isn't real. If we don't think about them (the animals), maybe they're not really somewhere out there, suffering. It certainly isn't something most people *want* to think about. And so, we actively *unthink* it.

George Orwell, in his dystopic novel *1984*, calls this process *doublethink*. Orwell explains doublethink as the process of denying reality:

> To know and not to know, to be conscious of complete truthfulness while telling carefully constructed lies, to hold simultaneously two opinions which cancelled out, knowing them to be contradictory and believing in both of them, to use logic against logic, to repudiate morality while laying claim to it . . . to forget whatever it was necessary to forget, then to draw it back into memory again at the moment when it was needed, and then promptly to forget it again: and above all, to apply the same process to the process itself. That was the ultimate subtlety: consciously to induce unconsciousness, and then, once again, to become unconscious of the act of hypnosis you had just performed.[1]

Doublethinking should be familiar to us because we do it all the time. Every time we engage in ignoring or denying something, we engage in doublethink. In order to ignore or deny something, we first have to know that it's there and, in some measure, know what it is.[2] Both ignorance and denial presuppose an awareness of what there is to ignore or deny. Ignoring someone at a social gathering presupposes knowing they're there and knowing which one they are. Denying (or being in denial about) the alcoholism of a spouse requires recognizing their alcoholic behavior so that you can deny what it is by interpreting it as something else. In doublethinking, the first thought recognizes the truth or existence of what the second thought denies; the first thought ensures the proper aim of the erasure accomplished in the second thought. And as Orwell points out, the erasure encompasses even itself in the forgetting that denies there was any ignoring, denying, or forgetting. But the truth or existence that is denied is not completely eliminated. It maintains its hold on us from the back of our minds, where truth and reality are routinely banished, and we have to repeat our "carefully constructed lies," as Orwell would say, to maintain a comfortable level of ignorance.

Consumer awareness of animal slaughter is a common site of doublethink. Most adults know—if they think about it for even a moment—that meat comes from dead animals and that animals are killed to produce that meat. This killing, of course, involves violence and a fundamental violation of the living animal. But consumers are expert at not thinking about this violence. Some people effectively forget to remember where meat comes from, or they defer any further recognition of the suffering entailed in its preparation/production as a commodity. The whole idea of humane slaughter emerges from doublethinking. Humane slaughter, as a concept, requires that the violence of slaughter is first acknowledged and then replaced with the idea that, by placing certain restrictions or by practicing slaughter in certain ways, the process can be made humane instead of violent. Legislative reforms sprung up around this idea in the form of the US federal Humane Methods of Slaughter Act: to enable consumers to feel better about the killing required in the production of meat, thus protecting the innocence of the concept of meat. This doublethink, of course, is integral to the widespread, largely unconscious consumption of meat.

The dairy industry, too, *relies* on doublethink. Instead of acknowledging the violence involved in dairy production (the artificial insemination, the separation of calves from the cows, the way the dairy industry fuels the veal and beef industries, the day-to-day exhaustion over time caused by excessive milking, and so on), consumers create fictions about its impacts on the animals: dairy cows are happy cows. Dairy farming is a benign process for the animals. Cows just naturally produce milk all the time. Cows *need* to be milked. If we didn't use cows for dairy, they wouldn't exist. Cows chose to be domesticated because it benefited them. These are all narratives that help to erase or obscure conscious awareness of the actual process of milk production.

Doublethink is a learned process. We aren't born doublethinking; we are taught to doublethink. We learn that certain animals (cows, pigs, chickens, turkeys, and others) are here to be eaten. Melanie Joy, in *Why We Love Dogs, Eat Pigs, and Wear Cows*, calls this dominant belief system "carnism": the belief that it is acceptable to eat certain species of animals. And this belief system gets perpetuated across generations, as children are taught these ways of thinking.

Children, of course, are inquisitive creatures and inevitably at some point are usually curious about where meat or dairy or eggs come from. Many children who live in communities alienated from food production practices will be shocked to find out that meat was a living animal—that chicken nuggets come from a chicken or that a hamburger comes from a "spent" cow used for dairy. But in my experience, it seems, parents avoid detailed conversations about these issues, opting instead for the vague nontraumatizing version: maybe that farmed animals live happy lives on Old McDonald's farm until the time comes when they are (abstractly and without detail) made into meat. Or parents might say, in response to a child's concern about the fact that animals are being eaten, that this is what those animals are here for. In other words, as Melanie Joy points out, the reasoning is often simply: *this is just the way things are.*

It's likely that this kind of doublethink—consumer doublethinking about dairy or meat or egg consumption—will be familiar. But what about people who are involved in the daily practices of farming animals? It might seem that they would be immune to doublethink, faced as they are with the practical realities of food production. However, through my research, I've found that farming requires a different mode of doublethink and, in turn, a particular form of training new generations to farm.

Elsa, the cow at Animal Place who had lived for fifteen years at a Waldorf farm school, is a reminder that forms of education in doublethink exist with the aim of teaching children not raised in farming communities to farm. These children are taught to understand the practices necessary to raise, breed, and slaughter animals for food. At the same time, they learn something deeper and more fundamental; they learn that farmed animals are here for us to use. This was echoed even in the Woodland Park Zoo educator's performance of farm education: "What do we use cows for?" The *use* of farmed animals is a taken-for-granted norm that gets reproduced in the education of children about these particular species.

4-H: EDUCATION ABOUT FARMED ANIMALS

Like Waldorf, other farming educational programs for children—such as 4-H—aim to educate and inspire a new generation of animal farmers. The 4-H organization is a global model of youth education; in the United States, one common feature of 4-H programs are their educational models dedicated to training a new generation of children to farm. The 4-H program is unique because it teaches children an intimate kind of animal agriculture—one where they bond closely with a single animal at a time and they learn how to give that animal excellent care.

When I returned from the California sanctuary trip, I wanted to understand the education materials commonly used in 4-H dairy education, and how these can affect participants' relationships to farmed animals. While an in-depth study of 4-H was beyond the scope of this particular project (and indeed, could be the subject of its own book), I did a textual analysis of the 4-H dairy curriculum, and I also wanted to talk with some people who had done 4-H as children. I chose to interview adults because I was most interested in how 4-H education is structured and how it was integral in shaping adults' relationships with farmed animal species. This is important because 4-H is designed to shape who a person is and how they relate ethically to humans and other animals; it is meant to instill a work ethic, an ethic of care, and an awareness of others. In this way, adults who completed 4-H are perhaps better equipped to comment on this transformation than children currently enrolled in the program, because they have the temporal awareness and reflective capabilities to comment on how it changed them. For a more in-depth study of 4-H, it would be desirable to interview both adults and children (as well as 4-H educators) in order to get a more robust picture of the program and its impacts. In the end, I supplemented my textual research with four interviews of adults who had done 4-H as children, one of whom I knew already. The others were people I was connected with via colleagues and research contacts.

The first person I interviewed about 4-H, as it turns out, was an animal studies colleague of mine—Allie Novak—who had done 4-H as a kid growing up in North Dakota, and she agreed to sit down and talk with me about her experience.[3] It's important to mention at the outset that Allie's background as an animal studies scholar means that she has spent much time in her professional life thinking about ethical and political questions about human-animal relations and that her graduate education and professional experience shapes how she thinks about and reflects on her 4-H experience.

We sat opposite each other in a coffee shop sipping some tea while she reflected on her experience as a child participating in 4-H along with most of the other children she knew. She began by telling me that she never raised a calf through the program, though she had raised other animals. Her older sister raised a steer named Teddy, to whom that sister and the family as a whole became very attached. As is the norm in 4-H, they eventually sold Teddy for slaughter on Achievement Day—the last day of the fair, when children auction their own animals as a conclusion to their 4-H experience, with a chance of winning ribbons for their achievements. When her family said goodbye to Teddy, Allie's mother said, "We're never doing that again." They found that raising steers for 4-H was difficult—the relationship was too close. They connected and bonded so closely with Teddy that it was heartbreaking to sell him for slaughter. Allie reflected that steers were somehow "too much of an animal," by which she meant that they had too much personality and were too much thinking, feeling beings.

But Allie felt this way about other animals, too. She raised her first 4-H animal at eight years old—a lamb she named Skittles. Allie and Skittles bonded closely over the summer they spent together. Skittles was housed at a nearby farm, and Allie would visit her daily, care for her, and take her for walks. Skittles followed Allie everywhere during these visits, and she remembers that she liked to show off how much

Skittles loved her. At the 4-H shows at the end of the summer, they would take naps in the barn together. That first year in 4-H, Allie didn't know what was coming at the end of the fair. As the summer's end neared, her parents realized how much she had bonded with Skittles and sat down with her to discuss different options. They considered giving Skittles to the local zoo, but the zoo said that they would eventually feed her to the lions, so they decided against that option. Ultimately, they decided that Skittles would be auctioned with the other 4-H farmed animals on Achievement Day. These auctions were social events and a way for the community to come together to support the children's 4-H projects. Allie told me that if a cute little girl were auctioning off an animal, the audience would intentionally bid up the animal so that the little girl would get a good price for the animal she raised.

Allie's eyes filled with tears as she told me about leading Skittles into the auction ring. She remembers it as an out-of-body experience; those moments in the pen were her last with her lamb, and they were a public spectacle. She remembers vividly the details of Skittles in the auction pen with the purple ribbon over her back, the steel gates of the pen, the cold gray day. She did not cry at the auction; she was very contained as she stood there selling her lamb. When she walked out of the pen, she had to hand Skittles off to a stranger, and they did not see each other again.

Allie looked down into her teacup and then, when she looked back up at me, said, "We were bonded. And I betrayed that . . . I betrayed that." She paused before continuing: "The lesson I learned that first time in 4-H was that you don't get close to your animal, and this happens to most kids the first time. But you learn, and you don't make that relationship again. In that sense," Allie reflected, "4-H is a lesson in the proper emotional relationship between humans and animals."

Through doing 4-H, Allie said, "I gained a sense of interdependency and felt like I was connected to other beings, but at the end of the day I learned the appropriate relationship." This shift did not change the way she physically cared for her future 4-H animals (she continued to give them excellent physical care), but it radically changed her emotional relationships with them. And so, the mode of doublethink in 4-H involves an emotional and intellectual acknowledgment that the animal is an individual with a unique personality and life of her own, and then that reality is quickly eclipsed by the narrative that the use of the animal for food is the ultimate aim.

As we chatted, Allie was ambivalent about the value of 4-H. She said, "The people who do 4-H are generally good folks and very ethical. They are responsible, caring, self-aware, and responsible to others. And so, on one hand, the experience teaches you that the animal has a face and that there is an ethic of care involved in farming animals in this way. But this is eclipsed by another kind of teaching—an education that reinforces the human-animal divide. 4-H is also an anachronistic education—you're being taught a form of farming that doesn't exist anymore, and the nostalgia for a different time and mode of farming is what keeps 4-H going. For me, it was this nostalgic salt-of-the-earth appeal that made me want to do 4-H in the first place. It made me feel like I was doing something real and important."

Historically, 4-H emerged in a climate of shifting rural-urban social relations. In the late 1800s, young people from farming families were leaving rural areas to seek urban employment. The growth of cities at the turn of the twentieth century, paired with the declining appeal of a lifetime of farm labor, caused great concern for the health and vitality of rural communities and the future of agriculture. Additionally, older farmers at this time were resistant to adopting new technologies from university agricultural science programs, and so youth were targeted as a way to help introduce new agricultural technologies to farmers across the country.[4] Youth were taught about new technologies and, in turn, taught those technologies to their older family members, who may have been more open to learning about them from their own children or grandchildren than from strangers.

As part of my research, I investigated the 4-H dairy curriculum because I wanted to understand how and what children are being taught about cows raised for dairy. The dairy curriculum is divided into three age groups: the *Cowabunga!* curriculum is appropriate for grades three through five; *Mooving Ahead* is designed for grades six through eight;

and *Rising to the Top* is targeted at grades nine through twelve. In the *Cowabunga!* curriculum, children are taught to identify dairy breeds, select a 4-H dairy calf, identify body parts, explore grooming and showmanship techniques, and learn about other basics of the anatomy and care of calves, heifers, and cows.[5] In the *Mooving Ahead* curriculum, youth are trained in judging and identifying show qualities in cows used for dairy, they are exposed to different career options related to dairy production, they are taught to engage in ethical decision making, and they learn the details of animal care in the dairy industry, including food, housing, parasite prevention and treatment, milking, and food safety.[6] *Rising to the Top* is the advanced program of dairy education for youth and involves learning detection and treatment for mastitis, detecting pregnancy and delivering calves, balancing food rations, selecting calves through records, promoting dairy products for sale, and exploring career options in greater depth.[7]

In the activity book *All about Dairy Cows* for the *Cowabunga!* curriculum, students are asked: "What kinds of cows make milk?"[8] The answer is a page of full-color photographs of the most common breeds of cows used for dairy: Holstein, Jersey, Brown Swiss, Milking Shorthorn, Guernsey, and Ayrshire. Of course, the intention of the lesson here is to teach about common dairy breeds and how to identify them. However, the way in which this question is posed suggests that not all cows produce milk—and that these particular breeds' inherent purpose is to make milk. Indeed, these breeds have been developed over generations to produce high volumes of milk that is extracted for commodity sale, but this question erases the complexities of the process of commodity milk production. One common myth about dairy that I heard throughout my research was that cows just "make milk." In casual conversations, I encountered numerous consumers of milk who were unaware that cows must have recently given birth to produce milk and that the body produces that milk *for the calf*. While this lack of awareness is odd, considering that most mammalian bodies that have recently given birth produce milk, the question "What kinds of cows make milk?" reinforces this misconception about *how* and *why* milk is made. The more accurate and relevant question for 4-H students would have been: "What kinds of cows produce the best milk for commodity sale?"

Other aspects of this same activity book reinforce the value of the cow as a commodity producer in less subtle ways. Children are asked "How much milk does a cow produce?" and are taught that "the average cow produces . . . 2,305 gallons of milk a year or about 8 gallons of milk every day of her milking period. That's enough for 128 people to have a glass of milk every day!"[9] These statistics on the productivity of the cow continue: "Each day the dairy cow can produce up to: 64 quarts of milk (256 glasses) or, 14 pounds of cheese or, 5 gallons of ice cream or, 6 pounds of butter."[10] The way milk is conceptualized as being measured in glasses of milk, pounds of cheese, gallons of ice cream, and pounds of butter reinforces the status of milk as a commodity good that can be bought and sold and consumed by humans, rather than its purpose as food for calves.

The bodily modifications that cows routinely experience are also made to seem benign: the page "Why do some cows look like they have earrings?" explains the utility of ear tags for cows and closes by asking the child, "Do you have pierced ears?" as a way to reassure the child that this process is benign—and connect it to the experience of getting pierced ears as a child, an experience that for many children is associated with excitement and a happy rite of passage.[11] Ultimately, these discourses and their role in the education of a new generation of potential dairy farmers work to naturalize the use of the cow and produce modes of doublethink that replace less savory understandings of these processes with cheerful alternates.

The second phase of the 4-H dairy curriculum, *Mooving Ahead*, focuses on developing ethical relationships with animals. Publicly, farmed animal shows have received some negative publicity over practices that constitute animal abuse—such as infusing udders with isobutane gas to make them look fuller and more engorged for show purposes.[12] Other common practices include injecting vegetable oil under the skin of the animal to give her a fuller appearance, or the use of clenbuterol, which causes weight gain in muscle rather than fat and is banned for use in farmed animals in the United States because it causes lung and heart problems in animals.[13] Exhibitors have been caught abusing animals, too, in order to enhance their appeal in the show ring. A graduate student in Oklahoma was caught beating a lamb to encourage swelling to give the lamb a firm feel to the judge's touch, and a pig was drowned by an exhibitor forcing the animal to drink water with a hose to meet the minimum weight requirements for a show.[14] These extreme examples are outnumbered by more common practices, like withholding food to lower weight prior to a show.[15]

Quality assurance programs and curricula involving ethical training are meant to improve these problems. This education is framed by a concern about the public image of animal agriculture: "With the passing of every show season such incidents [of animal abuse] cause animal agriculture to resonate more negatively in the eyes of the public. Perhaps the most serious consequence in the public arena has been decreased consumer confidence in the safety of dairy beef and milk."[16] Of particular interest here is the focus on improving public perception and consumer confidence in the industry—in other words, the focus is not primarily on the animal's welfare for the animal's sake. Animal welfare is a way to improve their marketability as a commodity and as a way to generate positive public consumer support not only for 4-H but also for dairy production and consumption more generally. Children are taught in this program that animals should be treated well not because they matter inherently as subjects of their own lives but because their treatment determines their value as commodities.[17] This is not to say that farmers, educators, and 4-H-ers don't care about the animals they raise; on the contrary, Allie's story illustrates the complex relationship of care involved in raising an animal for 4-H and the broader enactments of care in farming communities. Rather, it illustrates that care and the animal's welfare are constrained by the logic of economic commodification that understands the animal's welfare in terms of productivity and reproductivity.

The animal science education at 4-H, then, is a site where children learn doublethink. With the first animal they raise, 4-H-ers might see the animal for what she is: an individual with attachments, proclivities, a personality, and a social and emotional lifeworld of her own;

as a creature with an inner life, she becomes a companion, a friend. Using her for food production, though, involves learning to deny, ignore, and forget the complexity and depth of her being, which is displaced, thrown out of focus, overruled as irrelevant when her physical substance is commodified to represent a profitable return on investment.

The economic logics of animal agriculture are also a key point of education for children in 4-H. I interviewed Heidi Sloane, a twenty-year-old student at the University of Washington who had been involved with 4-H in western Washington for much of her adolescence.[18] I was connected to Heidi through a colleague and did not know her prior to our interview; in contrast to Allie, she had not taken any animal studies classes nor did she have an animal studies or animal rights background. Growing up in the suburbs of the Puget Sound region, outside of Seattle, Heidi's family was not a family of farmers; they lived in a suburban (somewhat rural) area and their house sat on a couple of acres that allowed them to keep several farmed animals at a time. Over the years, Heidi raised different species of farmed animals for 4-H.

"None of my school friends did 4-H," she said. "It was more something I did separately, outside of school, and I met a bunch of new people doing it."

Heidi was enthusiastic about 4-H: "My favorite animals were the alpacas. Once I got an alpaca, I didn't go back. Some of the fairs have fiber animal shows, and I started showing alpacas. They make the softest fiber. You can card and spin it and make things out of it or sell it. I got real into doing all that. My parents still have a couple of my alpacas out at their house."

I asked Heidi to share what she learned doing 4-H over the years.

"Oh, I learned a ton. The most important thing, though, was responsibility. I learned to be responsible for someone else's care. I love my alpacas and they live happy lives." Heidi paused and smiled, then added:

> And I also learned financial responsibility. Raising an animal and showing them costs money and taking on that responsibility teaches you how to manage money. You can also get cash

prizes at the fairs sometimes, and then you can also sell your animals. Food and bedding costs money—and [so does] vet care,
if they need it. So, you learn a lot about how to manage finances
and think about how money and the returns you get for putting
money into an animal are going to play out. I feel like I am a
much more responsible person financially from doing 4-H at an
early age.

Heidi had nothing negative to say about 4-H and was enthusiastic
about the value it added to her life and education.

Heidi focused on her alpacas over other species of animals she had
raised, and I wondered how 4-H experiences differed depending on
whether the farmed animal was raised for food or fiber. In Allie's case,
Skittles the lamb and Teddy the steer were raised for meat and had
to be sold at the end of their 4-H tenure; in Heidi's case, she kept the
same alpacas for years in a row, caring for them and showing them.
The relationships were different because of the designated purpose
of each species' role and involvement in 4-H but also because of how
their species was categorized.

I was interested in seeing how this played out in 4-H dairy shows
and so, at the end of that summer, Eric and I attended the Washington
State Fair. Each September, the fair is held at the Puyallup Fairgrounds,
about thirty-five miles southeast of Seattle, and, as in many places
around the United States, the Washington State Fair attracts people
from all over the state as a site of family fun and entertainment. There
are amusement park rides, arts and crafts, food vendors, concerts, a
rodeo, and animal shows. I had never been to the fair in Washington
before, but I had learned that 4-H had a major presence there, with
children showing the cows, goats, pigs, chickens, rabbits, alpacas, and
other animals they had raised as part of their 4-H education, and so
I wanted to attend. We went on the 4-H dairy day so that we could
watch the children show the cows they had raised for dairy. We arrived
early—just after the gates opened, hoping to wander around a bit before the 4-H shows began. Already, it was packed, as crowds of people
funneled into the fairgrounds.

CONSUMPTION AND CONSUMERISM AT
THE WASHINGTON STATE FAIR

One of the most noteworthy aspects of the Washington State Fair is how it is a site of consumption—of entertainment, of arts and crafts, and of food. And it was the consumption of food at the fair that was most striking to me. Although massive volumes of food are consumed and wasted in homes, restaurants, schools, hospitals, and offices daily, the enormous portions of food at the fair—half eaten and discarded, filling up waste bins that would then be dumped in dumpsters, picked up, and routed to the landfill—were staggering. Forty percent of edible food in the United States goes to the landfill, wasted, despite the fact that one in six people in the United States (i.e., 16%) are food insecure—in other words, living without a reliable daily source of food.[19] And although the majority of food waste actually occurs in people's homes, an event like the fair highlights the obscene amounts of food wasted in the United States because it is a site of conspicuous, public consumption.

With this knowledge of the paradox of waste and hunger nagging at the back of my mind, I felt overwhelmed by the amount of discarded food and excessive use of resources at the fairgrounds. And when I focused on what foods specifically were being consumed, I noticed that it was mostly animal-based foods: hot dogs and hamburgers, deep-fried frog legs, deep-fried macaroni and cheese, chocolate-covered bacon, sausages, barbequed ribs, fried chicken, ice cream cones melting into puddles in the trash cans, fish and chips, corndogs, scones with whipped butter and raspberry jam (a renowned Washington State Fair specialty), and so on. If consumption is the use that uses up, then consumption of animal foods is the use that uses up animal bodies and reproductive capacities, environmental systems, and the bodies of human laborers. And this was painfully obvious—if you paid attention—in the consumption and waste occurring at the fair.

It is easy to not think about waste. In many areas of the United States, consumers throw something in the trash or recycling bin, and it is picked up by a waste management service, gone from sight

and quickly forgotten. But when something is thrown out, it isn't eliminated; it goes *somewhere else*. And this "somewhere else" is often to
areas where communities with low incomes and communities of color
live, reflecting environmental racism, whereby those communities are
disproportionately exposed to health hazards from waste treatment,
landfills, and other industries that have devastating impacts on the
surrounding human populations (not to mention wildlife and natural
ecosystems).[20]

I thought about this as we wandered around, spending most of
our time in the portion of the fairgrounds dedicated to animal shows
and activities. We passed a corral containing a number of camels who
could be ridden for a fee. People climbed onto the backs of the camels
and smiled as the animals plodded around the enclosure, while family
members standing on the sidelines took photographs.

Not too far from the camel rides was a small pen housing a litter
of tiny piglets who were nursing from a giant sow. Fairgoers leaned
over the fence and cooed at the piglets who snuggled together, nap-
ping after their long nursing session. "Mommy, they're so cute!" a little
girl whined, "I want one!" And they *were* cute—unbelievably so. But
I wondered: what gets us from *they're so cute* to *I want one*? What gets
us from admiration or appreciation to a desire to own or consume?

Part of this desire is tied to the way we learn to be consumers. The
acceleration of consumerist culture has its origins in the 1800s, when
the Industrial Revolution made it possible for large quantities of commodities to be produced at lower costs than ever before. The increased
production led to increased consumption, and in the 1920s, for instance, General Motors aided in this acceleration when they began to
offer the option of purchasing a car on credit and, at the same time,
released new models of their vehicles every year to encourage the constant consumption of the latest model. After the Great Depression
and World War II, consumption boomed with the increasingly popular practice of purchasing on credit, paired with marketing campaigns
that encouraged people to want the latest, snazziest gadget. A kind of
manufacturing of desire was central to this process. Values assigned to
commodity goods changed, and the appreciation of the lasting quality

of a thing gave way to a desire to have the newest, shiniest thing. This consumerist culture, in turn, fuels the widespread reliance on credit and a heavily indebted population.

At some point during my adolescence, my parents decided it was time that my younger sister and I should learn to budget money. We were both given five envelopes and twenty dollars a month. We had to divide that twenty dollars among the envelopes in amounts determined by my mom: seven dollars to spending; five dollars to savings; four dollars to clothing; two dollars to gifts (for family and friends' birthdays, for example); and two dollars to giving (to a nonprofit organization). Over the years, my sister's envelopes were always overfull; she rarely spent money on anything and was good at tucking those dollars into her envelopes in her sock drawer and forgetting about them. My envelopes, in contrast, were *always* empty—all except the "giving" envelope (I felt too guilty to ransack the envelope of money designated for charitable donations). Otherwise, though, I spent from the envelopes indiscriminately. Money, even in my adolescent mind, was meant to be spent. And when my envelopes were empty, I would simply ask my sister to borrow five or ten or thirty bucks, and then I'd spend that too. Although my parents were both impressively frugal, master budgeters, who hoped to instill that quality into my sister and me, somehow, before I even knew what credit was, I was a consumer who spent beyond my means.

The culture of excessive consumerism in the United States and beyond frames the surrounding world as a site of consumption. Food, entertainment, and gadgets all become things to be consumed. Because animals are ownable, even live animals become objects of consumption. When animals are framed as objects (even thinking, feeling objects) to be consumed—when we can say "that animal is adorable; I want one," and then proceed to owning it—it is no wonder that the consumption of animals as food is so thoroughly normalized. This consumption relies on an understanding of the animal as an ownable object—whether dead or alive. The fact that pets or farmed animals are ownable—the fact that they can be bought and sold and killed without a second thought—sustains these living beings as objects of consumption.

We often learn this first with animals who are kept as pets in homes around the world. We discover that we can go to a pet store or a shelter or to the home of someone who has listed animals on Craigslist and pick out an animal who catches our fancy, contain them in a carrier or with a leash, take them away from their siblings or parents, and confine them in our homes as an object of (in the best cases) love, affection, and companionship. In the case of the exotic pet trade, geographer Rosemary-Claire Collard argues, in her forthcoming book *Animal Traffic*, that the wild lives of animals are transformed into commodity lives through a process of capture, which involves the severing of social and ecological ties—the infant spider monkey torn from his parent's dead body or the parrot chick snatched from her nest. This severing of familial and social ties occurs also in the context of domesticated animals. Breeding dogs or cats, for instance, often relies on the eventual sale or gifting of the inevitable puppies or kittens—resulting in the division of animal families and the disruption of established social relations among animals.

Although pets occupy a very special place in our lives, hearts, and families, our relationship with them—our haste in taking a single kitten or two away from the kitten-bearing cat, the routine of leaving a dog alone all day at home, or the uncritical way in which we confine a bird to a cage or a fish to a bowl—is not unrelated to how we relate to other animals.[21] Farmed animals, too, experience the severing of their family networks and close relationships: the piglets in the pen, for instance, soon to be taken from the sow and fattened up before they are killed for meat, or the calves in the dairy industry removed from the cows just hours after birth. These traumatic separations are routine side effects of our ownership and enclosure of domesticated animals. But doublethink enables the forgetting or avoidance of all of this.

The animal barns at the fair were filled with animals who had mostly experienced separation from their social networks: horses, cows, pigs, alpacas, goats, chickens, rabbits, among others. The children who cared for them—attempting to fill in the emotional needs of the animals—were often hanging out in the stalls with the animals or were nearby, talking to their friends and family. Whether humans can adequately fulfill the emotional needs of other species who have been removed from their social networks is unclear. At the fair, there were clear bonds between the children and the animals they were showing, but in most cases, even these bonds would be severed when the 4-H project was over and the animal was sold. And this prompted me to think about how our relationships with members of other species (especially domesticated species) might be imagined and practiced differently.

INTIMACY AND EMPATHY IN EDUCATION

The 4-H shows were held in large pavilions with bleachers for spectators to watch the children and animals moving around the ring. Eric and I sat there, close to several families who had kids showing animals in the show ring. The children dressed all in crisp white shirts and white pants with polished black boots and black belts. Their hair was neat and tidy, carefully combed and gelled or pulled back in ponytails or plaited in braids. The animals were gorgeous—spotlessly clean and in beautiful shape, clearly well-cared for and loved. They were each led by a child out into the ring in a line. The judges observed for all sorts of qualities in each pair: the cow's dairy character (which focuses on the angularity of the cow's body and the openness of the ribs), body capacity and frame (strength and rump angle and width for ease of calving), formation of the feet and legs, and (most important) the mammary system (specifically, the formation of the udders and teats).

We watched the children and animals in the ring, actively showing the animals, but we also observed the kids with their cows who stood at the far side of the ring waiting for their turn. One pair, in particular, caught my eye: a little boy who couldn't have been more than eight years old, with a small Guernsey heifer whose creamy brown and white coat was immaculately clean and shiny. The boy and the heifer stood out because of the way he continuously showered her with affection. He caressed her face with his small hands and kissed her on the nose,

on the side of her face, and on her ears. She licked him each time his
face came close to hers and nuzzled him with her nose and head. An
adult walked by and handed him an ear wipe and he gently wiped
out the heifer's ears to make sure they were spotless. Then he leaned
against her and rested his head on the side of her neck until it was time
for their show. From across the ring, I could see his lips moving, and
I realized he was talking to her (maybe whispering) the whole time
they waited. I wondered what he was saying.

When it was their turn, the boy held the lead tightly and led the little
heifer out into the ring with the others. They stood up straight and
walked around the ring, stopping periodically and turning in accor-
dance with the guidelines of the show. Several times, the heifer tried
to turn to nuzzle the boy and lick his face. He struggled to keep her
facing forward; her focus was on him and not on walking in line with
the other contestants. When it was time for the scoring, the pair did
not score highly in comparison to the others. The boy didn't seem to
mind; as they left the ring, he patted and kissed the heifer again.

As I sat there watching, I was moved by the clear love this boy and
the heifer shared. And, thinking back to Allie's story, I felt a rush of sad-
ness thinking about the lesson this boy would learn—and the trauma
of the impending separation for both of them. This boy, at the start
of his farming education, would learn that emotional attachments to
individual farmed animals were at odds with the logics that required
the animals' sale. He would have to push the bond he felt with this
heifer out of his mind as he aged in order to do the work of farming.
He would have to learn doublethink.

The 2009 documentary *Peaceable Kingdom: The Journey Home* features interviews with several people who were formerly involved in
farming animals. One of the farmers they interview, Harold Brown,
talks about his own background in farming. He was raised on a Michigan beef ranch and spent more than half his life working on farms that
raise animals for food. In the film, he talks about his involvement in
slaughtering animals, saying that the first time you kill an animal it's
really emotionally difficult, but you do it again and it's a little less difficult, and if you keep doing it, you eventually feel nothing. He said,

"The last thing you ever want to be is weak. Weak farmers don't survive." And so, he explains that you adapt and you close off your emotions to do the work. Doublethink becomes a mode of survival.

Learning how to relate to farmed animals seems to involve, then, a disavowal of the deep emotional connections farmers and others may feel with them—whether it is the personal connection with an individual animal (like Allie and Skittles) or the abstract appreciation and respect for farmed animal species (in other words, the way a child who has never met a cow or pig or chicken might feel empathy for them). Empathy—*feeling with* and *for* farmed animals—necessitates a different kind of relationship with them. As Lori Gruen argues in *Entangled Empathy*, empathy urges a response.[22] It is possible either to reject these empathetic responses in service of continuing the status quo or to respond by radically reimagining relationships with farmed animals. In other words, it is possible to identify and reject the tendency to doublethink.

Once Harold Brown let himself feel his emotional responses to farmed animals again, he was determined to change his relationship with them. Spending time at sanctuaries for formerly farmed animals, he got to know animals in a different context—one where he didn't have to engage in denial about the things he was feeling about them. In response, he founded an organization called Farm Kind that helps farmers raising animals for food to transition to other modes of plant-based farming.

How farmed animal species are conceptualized does not have to begin and end with their usefulness to humans as food sources. While I was at Animal Place, Marji Beach told me the story of how she found her way to working at the sanctuary, and it was a story that stayed with me—at the forefront of my mind—as I ruminated on how relating to farmed animals is learned.

Before coming to work for Animal Place, Marji got her degree in Animal Science at the University of California, Davis. She entered the program because she had always loved animals and wanted to figure out a way to work closely with them. As a kid, she thought the way many kids who love animals think—that the only way to have a career working with animals is to become a veterinarian. She wanted to work with farmed animals, so she sought out an animal science program where she could study large animal care and management.

Marji was working at the dairy farm on campus, learning the practical dimensions of dairy production. One of the cows was in labor, and everyone was standing around watching. The cow was having trouble giving birth. Her calf was stuck, and Marji was charged with helping to deliver the calf. Once Marji helped to maneuver the calf's body through the obstruction, he tumbled out in a cascade of amniotic fluid. When the calf stood up, he was taken away from the cow. The cow bellowed repeatedly, and the calf's higher-pitched call answered back. But they didn't see each other again.

This was a turning point for Marji. She felt empathy for the cow and calf and, listening to that emotional response, began to seriously question the education she was receiving and the practice of farming animals for food more generally. She said, "It's so funny because they tell you, they tell you everything there, you know, you learn all about debeaking and confinement systems and it's made to seem very normal and the way it should be and these professors have thirty plus years of research and experience and so you trust them. And so you know that the dairy calves are taken away right after birth. You know it objectively and rationally, but then when you see it, though, it's a whole different ball game. When you see the cow when she . . . I was actually the one who got to pull the calf out. She had dystocia—she was having trouble giving birth—and when they took him away, oh my god, she stood up and I had never heard an animal cry like that but you could feel it shake your bones. It was very powerful. And the calf started crying, and it was just awful. And I went vegan because there was my glass of milk right there and it was very disheartening to see that connection."

As a result of this experience, while Marji was still in the animal science program, she began to start advocating for farmed animals. "I saw how rough they had it," she said, "and how we're teaching our future generation—I mean we had three thousand students, and I was the only vegetarian and, you know, we're teaching them to be desensitized and detached from these really intelligent, emotional animals, and I didn't want to be a part of that anymore."

Animal-use industries, like animal agriculture, perpetuate powerful discourses about the role of animals in society and normalize violent practices, like confinement systems, dehorning, tail docking, and branding, among others. Dairy farming—and animal agriculture at large—*requires* engaging in doublethink, the more completely, the better. Individuals like Marji and Harold Brown have committed to a different kind of discourse, a way of thinking *and acting* that acknowledges and responds to their emotional bonds with other species with tangible changes in how they live and eat. Of course, this mental and emotional work is an uphill struggle against the power of doublethink and the industry's reliance on facilitating doublethink among consumers and producers. Nowhere did I feel more keenly the weight of this challenge—and the power of dairy doublethink than during my next two research trips: first to the World Dairy Expo in Madison, Wisconsin, and then to California's Central Valley to see the industrial-scale dairies.

8

"THE STAMP
OF DAIRYNESS"

In October 2012, I flew from Seattle to Minneapolis, stayed for a night with an old friend from college, and then drove the next day to Madison, Wisconsin, in a rental car to attend the World Dairy Expo. The interstate highways on which I drove passed through a classic northern midwestern agricultural landscape. A combination of forested hills and farmland, the landscape was dotted with cows on farms where the animals were pastured. Larger, more industrial indoor facilities were also occasionally visible, set back from the road. It was autumn and the leaves were changing, washing the landscape in a blaze of yellows, oranges, reds, and browns. The beauty of the drive—the picturesque view of black-and-white Holsteins grazing before an autumn backdrop—distracted me from what I was learning about the impacts of dairy production on the animals in the industry (pastured or not).

This was "America's Dairyland," and there was something wholesome about the landscape, the vision of what it meant to be an *all-American farmer* raising *all-American cows* for dairy. There was something nostalgic about the connection of this landscape with the production of milk, ice cream, and cheese—products that have been made integral to a US American diet and cultural traditions: childhood glee over an ice cream cone in summer, grilled cheese sandwiches, a glass of milk with every meal, and macaroni and cheese as a staple for many children growing up today in the United States. These foods, and the joy for many that comes from eating them, are wrapped up in this vision of farming—one where a few cows (not too many) graze the endless rolling hills and voluntarily return to the classic red barn for milking, making possible that tall glass of milk or that cheese and ice cream. Even as I was embedded in a research project that was uncovering the fiction of this vision of dairy farming and the distinct violence against the animal involved in the production of these food commodities, the aesthetic beauty of the landscape persisted in recalling this fantasy.

The World Dairy Expo is an international "cattle" and trade show with hundreds of vendors showcasing the latest dairy technologies and the top genetic heritage in animals used for dairy.[1] The expo hosts a show for each of the seven nationally recognized dairy breeds (Holstein, Jersey, Brown Swiss, Guernsey, Red and White, Milking Shorthorn, and Ayrshire); each show chooses a grand champion, and then those seven champions compete for two best-in-show spots for the year. In addition to these main show events, there are also youth events where children involved with 4-H and Future Farmers of America compete for their own prizes, like at the Washington State Fair. While these shows are going on throughout the week, the rest of the expo grounds are dedicated to the exhibitors who have come for the trade show.

The expo grounds are expansive, with several large pavilion and arena buildings packed with exhibitors. The vision of the countryside from the interstate was one of low-tech, minimal human interference and family farming. The expo's vision of dairy farming was a futuristic, technologically advanced model of efficiency and profitability. I entered the expo grounds and immediately got swept up in the sea of people moving through the aisles of booths. The place was a maze of row after row of exhibitors, sprawling through multiple buildings.

One of the first things I saw was the latest state-of-the-art milking machine. It was set up in the center of one of the expo halls and shone with its bright new metal and informational signs advertising unique features of how efficiently and completely it can milk a cow. The dairy industry already operates almost entirely with the use of milking machines—even a relatively small operation like Ansel Farm used milking machines to facilitate milking twenty cows at a time to eliminate the need for as much human labor.

The growing trend, however, is toward even more fully automated systems, and robotic milking technologies were creating the latest buzz at the expo and beyond. These are systems that take over every phase of the milking process. The cow enters the robotic system on her own, incentivized to do so because the enclosure contains her food. A robotic arm then prepares her teats for milking, cleaning them and pre-milking them, before continuing on to milk each cow. If anything is out of sorts, the machine reports any irregularities through a computer program. If the cow lingers too long in the enclosure, the machine will encourage her to leave, often by delivering a mild electric shock.

As I walked around and around the milking machine models, I wondered about the long-term impacts of increasingly mechanized production—especially when there were living beings involved. The technological treadmill ignited by the Industrial Revolution is defined by the constant need to innovate—to design and implement newer, better, faster technologies to replace the current model, which rapidly becomes obsolete. Milking machines and other dairy production technologies focus on maximizing the extraction of high volumes of milk as a commodity good. The animals themselves also become sites of technological innovation, bred for increasingly productive bodies that can generate dramatically increased volumes of milk when compared to the production capacities of their ancestors. These modern marvels of technological innovation in industrialized capitalist food production were on display at the expo.

Another emerging technology in dairy farming is embryo transfer technology, and while displays of it were not prevalent at the World Dairy Expo when I was there, it is poised to be a major new development in reproductive technologies. Embryo transfer technology involves hormone treatment for the cow (usually those with what the industry terms *superior genetic heritage*) and then collection of multiple embryos. The embryos are then either transferred directly into a surrogate cow or they are frozen and sold for later use. Embryo transfers are marketed as a way to substantially increase productivity and efficiency in the dairy industry. As Sara Kober of Trans Ova Genetics explains, "Embryo transfer [ET] is one option that can increase a cow's reproductive efficiency, allowing her to have numerous calves per year. While the average cow produces six to seven calves in her lifetime, ET can increase her reproductive efficiency to numerous calves per year—allowing breeders to multiply the success of their superior pedigrees."[2] Embryo transfers, then, represent a next step in increasing the incursion into cows' reproductive lives and eking out more reproductive material from each cow's body. Methods of reproduction like these pose interesting questions about the role of technology in dairy production, and the expo is a site where technological innovations are centered as desirable developments in farming practices.

At the expo, I immediately realized I would have to be systematic if I were going to see everything. Row by row, and building by building, I stopped at every booth to see what products were being advertised. There were companies selling bull semen, calf formula and flavor additives for calf formula, compost and waste management systems, milking systems, barns and housing, mats for cows to lie down on in the barn, ear tags, branding and castration tools, vaccinations, transport trailers, udder ointments, calf hutches, and so on. I watched public presentations about the latest technologies and I collected hundreds of pieces of free literature advertising the products on offer at the expo. I chatted casually with exhibitors and listened as they explained the benefits of their products.

One booth I visited early in the day was selling "heat detection" stickers (stickers that determine when a heifer or cow is in estrus). There were two men tending the booth. Before they spoke, I could feel them watching me, looking me up and down, as I perused the literature on offer and tucked several leaflets into my bag. Their gaze made me feel self-conscious. There were sample stickers laid out on the table, some with color exposed to show how easy it would be to detect in a herd of cows. I asked about the stickers and how they worked. Both men cracked a smile and glanced at each other. Then, one of them leaned in on the table, moving his face close to mine, putting his hand on my shoulder. He picked up one of the stickers and said, "Well, honey, when an animal mounts the heifer or the cow, that creates friction, and that friction rubs the sticker." He rubbed my shoulder, and

I inched away slightly, as he kept talking: "And that's what reveals the color underneath." He maintained eye contact and shuffled around to the front of the table to maintain his too-intimate grip on my shoulder, his hand moving down onto my back. I nodded, feeling uncomfortable. "This way," he continued, "we use nature to tell us when a cow's ready—you know, when she's coming into estrus. The other animals have a way of detecting these things, so we use that skill to our advantage. These stickers help keep track of all that." I pulled away, then, and thanked him as I moved to the next booth, feeling unsettled by the overly familiar way with which he insisted on touching me.

At several other semen supply and reproductive technology booths, I had a similar experience with men leaning in close to me, confiding in me about the products, touching my shoulder, arm, the small of my back, and making sustained eye contact—qualities of an effective salesperson but with an undertone of something else: a gaze that lingered too long as they looked me up and down—an odd half-smile as they explained enthusiastically, in explicit detail, the reproductive process. These interactions were gendered in an obvious way—it became clear immediately that it was unusual for a "city girl," as I was called on more than one occasion at the expo, to be interested in the reproductive process of cows raised for dairy.

"Why's a city girl like you interested in bull semen?" one of the semen salespeople laughed nervously.

"How could you not be?" I joked, and then added, "I'm doing a research project on dairy production and I'm interested in learning about the entire reproductive process." As I talked with various semen suppliers, I learned about the role of the bull in the reproductive process and the role of animals deemed by the industry to be male in the dairy industry more generally.

THE LIFE OF THE BULL

When a calf determined to be "male" is born in the dairy industry, he is destined for one of several futures. The overwhelming majority of male calves in the dairy industry are slaughtered shortly after birth for veal or they are raised up as steers for beef. But a small percentage of male calves with exceptional genetic heritage are raised, intact, for breeding purposes. These bulls are typically either used as what the industry calls "natural service bulls" or they are raised on separate breeding farms for semen extraction for use in artificial insemination. Bulls are reproductively mature at fourteen months of age, and it is preferable to use bulls who are young—2 to 2.5 years old—who are at their prime of fertility and are less likely to carry sexually transmitted infections.[3]

Natural service breeding involves the consideration of many factors, including "semen quality, libido, mating ability, and social ranking among other bulls and females."[4] In order to ensure the continued vitality of a potential breeding bull, a "breeding soundness evaluation" is recommended on a regular basis.[5] This test is meant to ensure prime breeding abilities and is also integral to identifying sexually transmitted infections in herds. Sexually transmitted infections are fairly common in dairy herds, and a combination of testing and vaccination is recommended to maintain healthy bulls and cows.[6]

When bulls are no longer viable candidates for breeding purposes in natural service herds, they are sent to slaughter. Healthy bulls with what are characterized as bad temperaments are also routinely culled from herds.[7] Temple Grandin notes how the separation of animals that is common in the industry influences bulls' temperaments: "If the bull calf has been raised alone and has not had the opportunity to interact with other cattle, he thinks he is a person and he wants to exert his dominance over the 'herd.' . . . Bull calves raised on a cow [i.e., nursed until weaning] were the least likely to attack. When they are raised with their own kind, they know who they are and they are less likely to think that people are part of the herd."[8] Another effective method for influencing the temperament of bull calves is early castration, which is, of course, impossible if an animal is to be used for breeding.

Castration, when it is performed, is usually done before the animal is six months of age, often earlier. Males are castrated for a number of economic reasons important to the industry, in spite of the fact that the procedure causes slow growth rate and compromises weight gain.[9] Castration lowers testosterone levels, reducing sexual activity and aggressiveness, making them more appealing to buyers who would raise them for meat.[10] Castration also lowers the muscle pH of male animals, which affects the quality and taste of their flesh.[11] Consumers in the United States are accustomed to the flavor and quality of flesh from castrated animals.

There are several types of castration used in the industry (physical, chemical, and hormonal), but physical removal of the testicles is the most common. Physical castration involves using a knife, like the Newberry castrating knife, to slice open the scrotum, after which the testicles are removed using what's called an emasculator tool (or the farmer's thumb). Once the testicles have been removed, the farmer will generally apply an antiseptic powder or spray to the scrotum to prevent infection. A second surgical method of castration is to cut off the lower third of the scrotum and pull the testicles out and sever them or pull them until the cord attaching them breaks.[12] Other common methods include the use of an elastic rubber ring applied at the base of the scrotum to restrict blood supply, after which the testicles take about three weeks to shrivel and fall off, or the use of a burdizzo (also called an emasculatome) to clamp and sever the blood vessels leading into the testicles, rendering them nonfunctional.[13] These methods all cause some combination of acute and chronic pain, and the industry norm is still not to provide anesthetic.[14]

Bulls not castrated who show signs of resistance are characterized as having a bad temperament, and, thus, the industry creates more docile bodies through killing animals who do not fit the particular standard of a "good" (read: *nonresistant*) disposition. Interestingly, the discourses that promote the bull as a virile, masculine figure, which I explore later, below, are partially at odds with the practice of culling bulls with "bad temperaments": despite the prizing of bulls for certain features of masculinity, those who are *too* masculine or aggressive and have too much of a mind of their own—the bull who, in Temple Grandin's formulation, "thinks he is a person"—do not remain in the industry.

You might be wondering, then, how strong, willful, and sometimes aggressive bulls are managed in the industry, since they are typically well over a thousand pounds and one or more two hundred–pound farmers have to control them. One method is the use of the electric prod, a wand of varying length that delivers an electric shock to the animal to keep him moving in the direction the farmer desires. Another mode of control is the nose ring, which involves a large metal ring being pierced through the bull's septum as a permanent bodily modification. It works as a method of physical control and dominance over the bull because the septum is a sensitive area and bulls are highly reactive to this sensitivity. Thus, a farmworker, who may be just a fraction of the bull's size, can control the animal through an exercise of power derived from causing pain to this sensitive part of the bull's body. These nose rings and the use of electric prods help farmers to facilitate the movement and control of bulls as the bulls are used in spaces of semen production.

Although some farmers choose to use the natural service method of reproduction on dairy farms, artificial insemination is increasingly the preferred method of reproduction because of its reliability and the level of control over the reproductive process it affords. Approximately 80 percent of cows in the dairy industry in North America and Europe are impregnated via artificial insemination.[15] Therefore, to speak about dairy reproduction is to speak primarily about artificial insemination. The advantages of artificial insemination include the ability for farmers to use semen from bulls with "superior genetic merit," to reduce the number of bulls kept on the dairy farm, and to sustain shorter calving intervals "resulting in a more consistent, uniform calf crop."[16] Note the mention of consistency here and the term *crop*, both references to the standardization of animals-as-commodities in agriculture (as though they were rows of corn or wheat).

In artificial insemination, the bull is typically not present on the dairy farm. Instead, bulls are increasingly raised on separate breeding farms. When raised for semen production, bulls are forcibly ejaculated two to three days per week and two to three times each collection day, using either an artificial vagina or an electro-ejaculator.[17] When an artificial vagina (AV) is used, a trained steer (a castrated male), a female in estrus, or a dummy is enlisted as a "teaser" to arouse the bull. As described in a University of Idaho teaching manual, the steps are as follows:

1. Mount—female in estrus, teaser animal or dummy
2. Restrain mount
3. Clean bull's sheath and belly
4. Lead bull to mount to tease and be teased
5. As bull mounts, grasp sheath and direct penis into AV
6. Hold AV near buttocks parallel to angle of vagina
7. Let bull serve the AV (don't thrust AV on penis)
8. Do not touch penis[18]

Teaser steers are commonly used and are made to stand in place while the bull attempts to mount him. Once the penis is erect, after several false mounts, the human handler manually diverts the penis into the AV. The AV is a long, tube-shaped device that facilitates ejaculation through "thermal and mechanical stimulation."[19] The AV collects the semen, which is then processed for use—typically stored and frozen for later sale. The use of an AV can be difficult and potentially dangerous for the human handler as well as for the steer and female teaser, either of whom can be injured by the bull's attempts to mount.

The other widely used method for semen collection is a process called electro-ejaculation, which involves a human handler inserting an electric probe into the rectum of a bull who has been restrained and delivering a series of electric currents to the prostate, causing the bull to ejaculate involuntarily.[20] This method is often more efficient because it is less dependent on the bull's willingness or arousal. However, this method often produces less effective semen than other methods, like the artificial vagina. Electro-ejaculation is also a painful procedure. In fact, in a number of other species, electro-ejaculation is performed under anesthesia, but bulls experience this practice without anesthesia.

That bulls are being housed on breeding farms specializing in semen production is the result of industry division and consolidation in a quest for greater efficiency and capital accumulation. One might not think right away about the bull's role in the reproductive process of dairy production—indeed, how often does one even think of dairy production as part of a reproductive process? And those who do imagine it may be unlikely to imagine a process like electro-ejaculation or even artificial insemination. However it is obtained, semen is an internationally traded commodity rolled into the worldwide production of other commodities—milk and meat. Semen and the bull himself are integral to the continued functioning of the industry as a whole, just as the pregnancy of the cow and the birthing of calves are indispensable processes on which the industry relies.

INDUSTRY ADVERTISING AS DISCOURSE

As I walked through the World Dairy Expo, I gathered great quantities of advertising materials: flyers, pamphlets, sample products, catalogs, and industry newspapers. By the end of the day, my shoulders and back ached, weighed down as they were by numerous tote bags (some mine and some given to me by industry reps like a fire-engine-red tote bag with Select Sires' logo on it) filled with free stuff from the expo. Later that night, exhausted in my motel room, I spread all of the materials I had gotten out over the two double beds and tried to make some sense of what was there. There was an overwhelming amount of information and ads for products I didn't know existed such as the powdered flavor additive for milk replacers that would encourage calves to drink rehydrated powdered milk from a bottle instead of from the cow's teat.

Once I returned home, I set to work in earnest trying to categorize and track the themes that threaded through the industry ads. One of the themes that repeated itself across the industry literature I collected was the gendered use of the animal and the way this use reflected human norms about reproduction, the body, and gender, with a particular emphasis on fertility and pregnancy. Another theme that continually emerged involved an appeal to US patriotism in the context of celebrating settler-colonial histories and contemporary militarism. Also evident was the pervasive humor about the highly sexualized nature of the dairy and semen industries. It is important to note that these industries frame animals through a binary understanding of sex and gender, categorizing them as being female or male and as reproductively viable or not.[21]

Why consider ads as an important artifact of analysis? Discourses that are circulated through advertising reveal key insights into norms about the body, gender, and animals that work to sustain the material practices of the industry. Michel Foucault argues that discourses reveal the way meaning is made using language and the construction of "commonsense" narratives about a particular practice or set of social relations.[22] These narratives do more than just circulate as stories: they affect the material ways in which humans interact with each other and with other species, and they shape the power inequalities within these sets of relations. Indeed, as Donna Haraway explains, "Discourses are not only social products, they have fundamental social effects. . . . Scientific discourses both bound and generate conditions of daily life for millions."[23] The discourses circulated about animals, gendered bodies, and reproduction that populate these dairy and semen industry materials are so illuminating that it is worthwhile to take some time to enumerate a series of examples.

Fertility and Pregnancy

That the industry advertisements at the World Dairy Expo showed a preoccupation with the cow's fertility makes sense since the industry relies on the continued fertility of a large population of cows. A cow's continued survival, as I've shown in previous chapters, is primarily dependent on her fertility. When her fertility declines, or if she is sterile to begin with, she is slaughtered. Because cows must give birth to calves at regular intervals in order to produce milk, their production of milk as a commodity hinges on their fertility. Consequently, industry discourses about fertility overwhelmingly link fertility to profit for the farmer.

To aid in the reproductive management program on the farm, the SmartDairy Activity Module by BouMatic tracks individual cows in a herd.[24] This technology attached to individual cows enables farmers to track sick cows, cows in various stages of pregnancy, and cows who are ready to be impregnated. A BouMatic ad says, "Find them. Breed them. Improve your profitability." Again, this links breeding and fertility to profit and reveals the economic priorities of the industry. Rovimix Beta-Carotene, a cow fertility solution created by the science-based company DSM, promises to increase farmers' profitability, emphasizing that "cow fertility problems are one of the costliest production issues in dairy farming."[25] The slogan for this product is: "We've conceived a better way to fertility." This statement acknowledges humans' involvement in the reproductive process—in fact, so much so that it suggests a kind of "playing God" scenario. This company has transcended the base, mundane path to fertility; they have "*conceived a better* way to fertility" (emphasis added). This cooptation of fertility and the reproductive process of cows and the relationship between fertility and profit link intimate politics of reproduction with the logic of the global flow of capital and economic efficiency. The fertility of the cow is directly tied to the fertility of the farm as a business. The ad promises that the product can "enhance your business growth." Seen as a form of fertility, the economic growth of the business is dependent on the cow's biological fertility and the growth and delivery of the fetus. The accumulation of capital in dairy production is tied closely to the reproduction of the cow.

Another tracking system—Semex—helps to monitor sick cows and to maintain an insemination schedule. With the slogan "Listen to your cows and take control of your herd," Semex promises "reduced labor, drug and semen costs; increased pregnancy rates; more of herd confirmed pregnant; less time spent in headlocks, more time making milk; quick return on investment."[26] The emphasis here is on control, efficiency, and profitability. The slogan's advice that farmers "listen to [their] cows" suggests that this management and control of the cow's body is related to a close relationship between the cow and the farmer—that the farmer *listens to the cow's needs*. This suggests that the cow is an active agent in the process, that she has an opinion and is listened to, that she and the farmer, perhaps, have a heart-to-heart conversation in which she confides that she's not quite ready to get pregnant again and maybe they ought to wait until next month. Echoing this suggestion that the animal has a choice in the matter, one Cargill advertisement states: "She's in it to make milk. You're in it to make a living. We're in it to help."[27] The cow, though, is not involved in the decision-making process — she does not consent to being inseminated and she does not consent to having her calf taken away or her milk diverted from her calf into the marketplace. In fact, the logic of the industry is such that practices involving the management of the cow's reproductive cycle are driven by the most basic biological capacities. If a cow *can* get pregnant, then she will be inseminated, and her consent or agency does not enter the equation.

And indeed, this logic of the cow's reproductive capacities dictating the process is affirmed in stark terms in a series of advertisements for Bovi-Shield Gold by Pfizer, a combination antiviral vaccine.[28] The various ads in the series pose the question, "If she can't stay pregnant, what else will she do?" Each ad pictures a Holstein cow doing something unbelievable: sitting in the passenger seat of a fire engine, standing next to a hunter with a dead pheasant in her mouth, and being ridden with a saddle by a child. These images work to remind us that, of course, a cow is not a Dalmatian or a hunting dog or a horse. Further, the ad reminds us that the animal figures (both absent and present in the ads) — the Dalmatian, the hunting dog, the horse, and certainly the cow, whose job it is to "stay pregnant" — are each here to perform a service to humans. The ad continues, "Keep your cows pregnant and on the job. . . . Ask your veterinarian or Pfizer Animal Health representative how to protect her pregnancy, your reproductive program and your bottom line."[29]

And yet, cows in the dairy industry have their calves removed immediately after birth and so don't engage in the work of parenting — the industry's way of reinforcing the assertion that a cow doesn't have anything to do if she's not pregnant. The product she is responsible for producing is, first and foremost, milk and so the *pregnancy* (rather than the resulting calf) is her job, not the care of offspring. These ads, then, reinforce norms about female reproduction (for both humans and animals) and they also reinforce the supposed absurdity of an animal doing anything other than servicing humans in their predetermined way. The ad reminds us that it is ridiculous to imagine a cow doing anything other than producing milk. She is not just a cow, but a *dairy* cow, and her job is unequivocally to stay "pregnant and on the job."

Colonial Histories and US Patriotism

One of the things I learned about through my research was the role of the cow in the colonial project. Contrary to what many grade school curricula teach, the so-called settling (a deceptively innocuous term) of the US West was a violent process of genocidal killing of indigenous human communities, enclosure of the survivors on reservations, and a systematic erasure over centuries of Native cultural practices and ways of life. The settler-colonial project also killed certain species—like the bison—in droves, driving them to near extinction. Historical accounts of this process don't often include the role of the cow in this process. But the cow was integral to the colonial project. Fencing and land clearing were primary modes of seizure and enclosure of land and resources, and much of this was justified by needing to clear the land for ranching. Cows were physically used to occupy land and change the prairie ecosystems, displacing indigenous humans and native animal species from their home.[30]

The name of one of the Select Sires semen collections (composed of twenty-two "elite sires")—Superior Settler—celebrates rosier versions of these settler-colonial histories.[31] The advertisements show Holstein cows in the foreground, in front of western US landscapes as backdrops. In one, there is a mountain range in the distance behind the cow, and in others, there is pasture and fencing in the background. These images recall the settler-colonial project in the West—both in the prominence of distinctly western landscapes and in the vestiges of settler colonization: the fences, the pasture cleared of trees and diverse prairie grasses, and the cow herself. Thus, Superior Settler semen unintentionally gestures to the colonization of humans and animals under capitalist agricultural systems. More likely, though, the motivation behind calling up the settler-colonizer narrative in this advertisement was to recall a less critical version of US history. Settlers moving west are seen by many as part of a distinctly "American" and highly celebrated, heroic historical narrative. For many farmers, in fact, their historical lineage is linked back across generations to the original settler-colonizers, and this is a source of deep pride. But settlers and their descendants live, work, and farm on stolen land, a fact that gets obscured by patriotic discourses and the white-washed American Dream.

The series of advertisements from Bovi-Shield Gold by Pfizer described earlier reflects a certain kind of US patriotism and the role of the animal in this patriotism. The image of the cow being ridden by a young child is reminiscent of settler-era horseback riding and, in fact, the cow is wearing a western-style saddle, cushioned by a Native American printed blanket (a proud display of western settler history and the traces of the colonization of indigenous culture). The fact that the cow is being ridden by a child assures us that calling up this history is a harmless, unproblematic, and apolitical enactment of US patriotism. Similarly, the version of the ad where the cow stands behind a hunter holding a dead pheasant in her mouth also reminds us of a distinct kind of patriotism and western settler/rancher culture—namely, that a hunter and his dog (in this case, cow) are a particular manifestation of cross-generational US white masculinity.[32]

In this scenario, even though the purpose of the ad is to show the absurdity of trying to use a cow for anything other than dairy, it is revealing of the ongoing settler-colonization narrative in that the cow is again enlisted in the colonization of wild animal species. The story is further complicated, though, because the common pheasant was native to Asia, introduced in the United States for hunting, and is now one of the most hunted birds in the world.[33] The pheasant is yet another colonized body, displaced from its native land, introduced as a now-somewhat-wild animal, and then killed in the sport of hunting in which the cow is enlisted to help. The cow with the pheasant in her mouth and the hunter smiling self-assuredly illustrate the use of humor to avoid taking seriously the violence against the dead pheasants or the more widespread histories of violence against humans and other animal species.

Hunting and western horseback riding represent a particular kind of US nationalism, just as the cow-as-firefighter version of the ad recall the familiar heroism of the firefighter figure. Particularly in the post-9/11 climate, the firefighter, who may always have been a local heroic figure, rescuing children and puppies from burning buildings and acting as first responders for daily emergencies, has reached a new, elevated status as hero. The fire-fighter-as-patriot here recalls in the US imaginary those who lost their lives in the rescue efforts in the aftermath of the World Trade Center attacks and the zealous patriotism that emerged in that political climate.

Similarly, another advertisement—this one for a dewormer called Cydectin—connects the production of dairy and the care of the dairy herd to the Wounded Warrior Project. "Our troops deserve the very best this country has to offer. Thanks to your purchase, Cydectin®, the #1 pour-on cattle dewormer, is able to support the Wounded Warrior Project®, and show our heroes how much we appreciate their service."[34] This emphasis on military service and the service of fire fighters, and the overt discourses of heroism attached to these figures, implicitly links the US farmer figure to this performance of patriotic service. Particularly in the post-9/11 political climate, discourses of soldiers protecting "American freedom" abroad circulate to justify continued occupation in distant places around the globe, and the image of the firefighter protecting, rescuing, and serving the US public at home helps to maintain a narrative of heroic service and patriotism. The farmer and the deworming company supporting the Wounded Warriors Project position themselves as patriotic allies on the home front. In fact, the cow herself is a patriotic ally, providing her dutiful service as part of the home front forces.

The cow pictured in every one of the Bovi-Shield Gold ads is a Holstein—the distinctly black-and-white US breed and an iconic symbol of US dairy that reverberates through children's books, dairy product packaging, and the US imaginary. Dairy production and, importantly, dairy *consumption* are constructed as integral to what it means to be *American*. Indeed, when I was staying the night outside Madison, at several restaurants and coffee shops, I asked for soy milk

(instead of cow's milk) in my coffee and no cheese on my meals. At
more than one place, I received a look of disbelief and a friendly-but-
bewildered "What is *wrong* with you, girl?!" Not only was abstaining
from cheese and dairy outside the norm in Wisconsin—in America's Dairyland—abstaining from these products felt practically *un-
American* in this context.

Sexualized Humor and Violence

Social norms reinforce a gender binary, and in a trans-exclusionary
and biologically essentialist move, frame women as quintessentially
equipped for motherhood; at the same time, these norms frame wom-
en's bodies as sexualized objects in popular culture. What I sensed was
an implicit objectification of my body when I spoke with the men at
the expo as they looked me up and down, slowly, was echoed in more
explicit ways in the advertising materials I collected. In fact, it was the
norm for bodies deemed to be female—mostly cows—to be sexualized in the industry. Before I even entered the expo halls, I had seen on
the drive several large animal trailers with the "mud flap girl" silhouette
on the mud flaps (a woman with an hourglass figure, large breasts and
long hair blowing in the wind). After all I had witnessed, not much was
surprising me at this point in my research, but the extent of the objectification and sexualization of the body in dairy industry advertising
materials came as a shock.

The discourses in semen catalogs reveal certain gendered attitudes
in the industry toward animals. Bulls are commodified based on their
genetic makeup, the quality and virility of their semen, their appearance, and their reproductive potential. And bulls are also made to take
responsibility both for the production of genetically superior milk producers (future cows) and for any subsequent violation (artificial insemination) of their bodies.

Select Sires sells semen from show-quality bulls, and their catalogs
exhibit a range of attitudes about bulls and cows/heifers. In one catalog, a bull named Alexander "puts the stamp of dairyness on his daughters like no other," and Sanchez "makes them special—tall, dairy and

strong with beautiful udders."[35] GW Atwood is "the hottest bull to hit the type market in years.... He makes the kind you can have fun with," and Java breeds cows with "great rear udders and attractive rumps." Governor, who has "greatness in his genes," produces daughters with "youthful mammary systems that catch the eye and stand the test of time." These advertising discourses suggest that the bulls are virile and masculine—not only capable of prolific reproduction but also capable of breeding exceptional (attractive and productive) females.

Alexander putting "the stamp of dairyness on his daughters like no other" indicates a sense of ownership over future cows. The "stamp of dairyness" recalls the practice of branding, whereby a literal mark of ownership is burned into the animal's flesh. It also might be a reminder of the practice used by certain animals of "marking"—the way dogs and cats, for instance, mark territory with their urine. Alexander's daughters are, of course, literally owned by the farmer that bred them, but in this advertisement, the responsibility for the ownership (and all the violence that comes with being legally categorized as property) is shifted to Alexander, who gave them "the stamp of dairyness" in the first place.

Sanchez, GW Atwood, Java, and Governor are all enrolled in the reproduction of attractive and sexy females—from these bulls' genes, a farmer can expect to get "beautiful udders," "attractive rumps," and "eye-catching youthful mammary systems"—the kind of cow "you can have fun with." This language fetishizes the female body and calls up the cultural preoccupation with eternally youthful feminine bodies. The implicit discourse here, too, is that these bulls are gifting the farmer with the cows of their dreams—a gift, from one "man" to another—not unlike the long institutionalized practice of fathers "giving away" their daughters to future husbands (either for payment or as part of an ingrained cultural tradition)—an odd connection, given the involvement of the farmers in the reproductive process.

Reference to "daughters" also might reference a sentimental imaginary of the father-daughter relationship—of bulls (and by implicit reference, farmers, too) as fathers and family men. And yet, we know from the nature of the semen industry and the commodification of the male body in that industry that the bull will never meet his offspring. His daughters and sons will be born as a result of artificial insemination on distant farms. Even the cows impregnated with his semen are likely never to see their offspring beyond their first few hours of birth. The subtle reference to family through the father-daughter relationship is out of place in the dairy industry, where bovine family structures are fractured, animals are alienated from one another, and their destinies are determined exclusively by their value as commodities.

An emphasis on the udders and mammary systems as a fetishized trait of commodity production is expressed in these same catalog images of cows with engorged udders and tails pushed aside to display prominently the cow's vulva. Images like these emphasize the promise of excessive commodity production (after all, big udders are suggestive of high milk production). However, the images of the udders also reflect the popular fetishizing of large-breasted women, and the advertisements are reminiscent of familiar popular sexualized discourses about women. "Youthful mammary systems that catch the eye and stand the test of time," calls up a cultural preoccupation with perky breasts and eternally youthful female bodies—bodies that can maintain an attractive, youthful appearance while at the same time being productive milk-bearing mothers. "Great rear udders and attractive rumps" is reminiscent of the fetishization of women's "tits and ass" in popular culture. The image of the cow's exposed vulva is meant to show that her vagina is open and ready for business, similar to the way in which pornographic images showing women's genitalia are suggestive of this same message.[36] And finally, GW Atwood making "the kind you can have fun with" promises that these cows are more than productive machines, they are attractive, well-endowed, promiscuous, fun-loving females ready for whatever might be in store for them. These images and discourses call up Carol J. Adams's now-classic feminist analyses in *The Sexual Politics of Meat* as they do the work of sexualizing the animal body.

Bulls' bodies in the semen industry are fetishized as sexy icons of virility and masculinity. Some semen catalogs show portraits of bulls that emphasize their physical characteristics—height, stature, and, most notably, their highly visible genitalia. These images show visible penises and large testicles as commodified means of production in the sale and extraction of semen. The commodification of these body parts linked to the animal's sexual experience and reproductive capabilities echoes the commodification of the cows' udders and vaginas as selling points. These pictures recall the important reproductive function of these parts in commodity production (milk, calves, semen). And just as the exaggerated visibility of cows' engorged udders and exposed vaginas is reminiscent of pornographic images, so, too, are the bull's penis and testicles suggestive of the human male erection and sexualized imagery fetishizing the body.

Some discourses present in the semen supply industry use sexual humor to obscure what seems to be a discomfort with the work of semen extraction and artificial insemination. Universal Semen Sales is an online semen supplier that also sells merchandise to advertise its products and services. Their mascot is a cartoon bull, named Sammy Semen, who walks on his hind legs and carries a small suitcase labeled "A.I."[37] The company sells boxer shorts with a large white cartoon smiling sperm cell and the words "SAMMY SEMEN" across the back. One t-shirt for sale has the large cartoon smiling sperm cell and beneath it, the text, www.universalsemensales.*cum* (emphasis added). A mug for sale has the company's Sammy Semen logo on one side and the words "Collection Cup" on the other.

Each of these products brings humor to the practice of semen production, but one t-shirt in particular is worth analyzing in some detail. The shirt shows Sammy Semen in the foreground, walking on his hind legs and carrying his AI case. He is sauntering up behind two smiling cows wearing bright red lipstick and with large udders and backsides angled at Sammy. The cows are eager; they are thrilled to see Sammy and his AI case. The implication is that artificial insemination is pleasurable—like good sex—and the cows want it. The company's slogan frames the cartoon scene: "We stand behind every cow we service: Universal Semen Sales." The slogan serves two important purposes. Primarily, it reassures the buyer that the semen is high quality and that the company stands behind the quality of their product. But the second purpose of the slogan is its undercurrent of sexual humor. In a practical sense, the farmer stands behind the cow as he inserts one hand into her rectum and the other into her vagina to perform the artificial insemination. The use of the word *service* intentionally recalls the colloquial use of the term *service* in a sexual context, which is to perform a sexual act for their express pleasure. Again, the cartoon and slogan suggest that cows want sex. Interestingly, it suggests that artificial insemination is an act of sex and, even as the responsibility is shifted to Sammy, we all know that it is the farmers (or at least their hands and arms) who engage in this act of sex with the cows, a point that is particularly noteworthy in the context of cultural discomfort with bestiality. This characterization of artificial insemination as a sex act is particularly odd considering efforts in other agricultural reproductive industries to characterize artificial insemination as a purely scientific practice.

And yet, this t-shirt also reveals industry discomfort with interspecies sex acts. Positioning Sammy Semen in the cartoon, holding the case, works to shift the responsibility of the violation of the cow away from the humans involved. It attempts to naturalize the process of reproduction and reassure us that cows are eager and excited about it. We are encouraged through this vision of artificial insemination to ignore the decidedly unnatural characteristics of industry procedure (e.g., human involvement in semen extraction and artificial insemination, the fact that the bull and the cow are no longer typically housed on the same farm, etc.). The joke of this cartoon is that bulls do not actually walk on two legs and carry the tools for artificial insemination. This image is humorous specifically because of its absurdity. But it also reveals the absent figure of the human farmer—the bull walking upright on two legs and carrying the AI case—these are distinctly human characteristics in this context. Even as the human involvement is revealed, the cartoon jokingly deflects the responsibility to Sammy. His presence there is essential to the fictional narrative because it is reassuring to think about the bull and cow as engaged in a "natural" process—and this is not entirely fictional: after all, it is *his* semen being used in the reproductive process.

It is also worth thinking about the target audience for this merchandise. Presumably the t-shirt was designed for farmers (primarily men) involved in daily practices of semen production and artificial insemination. This t-shirt was likely not designed with the general public in mind. Thus, customers who would buy this shirt (or who would be given it as a "thank you" for doing business with the semen farm) and find it funny would already know the realities of the reproductive process in the semen, dairy, and beef industries. They would know the absurdity of the image of Sammy Semen with the AI case. There is something nostalgic about the insertion of Sammy into this fictional scenario—there is a certain comfort in the imaginary of a different time and place where bulls and cows actually engaged in sex to reproduce.

Read as an acknowledgment of a nostalgic narrative (albeit a fictional one) of dairy production, this image of the cartoon cows in the field and Sammy in the foreground recalls the idyllic vision of dairy farming described at the beginning of this chapter. There is something romantic about envisioning cows on pasture in the rolling hills of the United States and the imaginary of cow families living together in these places, involved in a "natural" cycle of reproduction whereby cows willingly share a portion of their milk with the farmer and, eventually, the consumer.

* * *

As I drove along the highway to Minneapolis after the expo, I saw a sign on the road for a Laura Ingalls Wilder museum. As a child, I had read and loved Wilder's books that recounted her life in the US West. The first book in the series, *Little House in the Big Woods*, described her early life living in a log cabin in the Wisconsin forest. On a whim, I took the detour to visit the museum and ended up in a tiny town, Pepin, Wisconsin. After visiting the museum, I drove just outside of Pepin to visit the site where the "Little House in the Big Woods" had stood. Now there is a replica cabin and the cabin stands, not in a forest, but on a hillside in the middle of farmland. All around, acres and acres of corn grew and the landscape as far as I could see was farmland with only a few stands of trees remaining. I stood there on the hill by the Wilder cabin replica and thought about the development of the West, the wholesale clearing of land for ranching and farming, and the role of the cow and the bull in how the land was transformed. Discussing settler colonialism and the destruction of the prairies, Eli Clare writes:

> White colonial settlers claimed the land as their own, dividing it into neat rectangles, fencing it, and establishing herds of cattle. The near eradication of the prairies started here. The grazing and migration patterns of bison had been integral parts of these ecosystems, whereas cows destroyed the grasses, giving nothing back. And then white farmers literally tore up the prairie with their plows. They planted monocultures of wheat, corn, and soybean. One hundred seventy million acres of tallgrass prairie used to exist in North America; seven million are left now. Today when we eat corn or steak produced on agribusiness farms in the Great Plains, we are connected all the way back to that mountain of skulls. Monocultures start with violence, removal, and eradication.[38]

The picturesque landscape of "America's Dairyland," with cows grazing and rolling hills of pasture, erased and elided the violent appropriation, first, of the cows' and bulls' bodies for use in agriculture and, then, their use in the displacement of Native people and animals and the clearing of forests for farming. This idyllic landscape is, in fact, a testament to the historical colonization project and to the current deployment of that history to erase the violence of dairy production and ongoing erasure of Native communities and cultures. In spite of its peaceful veneer, this was a place of violence.

The realization that things you once celebrated and appreciated are actually linked to such violence is a difficult thing to stomach. Seeing the violence of dairy production, for instance, makes some people see eating in a different light. I have heard many ethical vegans talk about how overwhelming it is when they first go vegan and are thinking so actively about the violence embedded in the production of animal products to see the scale of consumption around them. What was once obscured because of its routine nature becomes highly visible and emotionally provocative when you experience it in this different light.

The continuation of animal agriculture—motivated by the demand for dairy, meat, and eggs, for corn as feed for farmed animals, and for cleared land for animal agriculture—is wholly dependent on the reproductive process. In order to sustain and grow the current market for farmed animal products and the institution of farming animals at large, animals must reproduce. Cows must be impregnated on an annual schedule if they are going to continue to produce high volumes of milk for commodity sale. Animals must be bred, born, raised, and slaughtered if the meat industry is to sustain itself. These are simply the realities of the industry. The reproductive politics of animal agriculture cannot be separated from its continued existence, and the materials at the expo emphasize this point. The juxtaposition of the bucolic rural landscape of northern midwestern farmland with the intensely capitalist place of the World Dairy Expo illustrates the tensions between the persistent public imaginary of US dairy farming and the economic logic of industry practices driven by commodity production.

As I flew home to Seattle, I thought about what more I needed to see as part of this research. One thing I came away from the expo thinking about was the scale of production, and the emphasis on growth and technological innovation promoted as the path forward. Up until that point, I had focused mostly on mid- and small-scale dairy farming, but the expo made it clear that the industry is heading in the direction of bigger, more mechanized production practices. I decided I needed to go back to California, this time to see the scale of production where farms house three thousand, ten thousand, fifteen thousand cows— the kind of place where Sadie was born.

9

CALIFORNIA DREAMING

When I returned home from Wisconsin, I booked a flight to San Francisco for my final research trip. I wanted to attend a larger-scale auction and see the massively industrial scale of dairy farming that is less common in the Puget Sound region of Washington. Dairy production in the United States has been moving westward from the Midwest, and California now produces one-fifth of dairy nationally. As an agricultural landscape, there is no doubt that California is an important site. This is literally where dairy is headed—both westward and toward increasingly mechanized, high-tech farms.

I arrived in San Francisco and stayed with Tish, who had attended the cull market auction with me in Washington. I rented a car at the airport on my way into town and we drove out from the Bay Area into the San Joaquin Valley, which is part of the larger Central Valley. California's Central Valley is one of the most productive agricultural landscapes in the world, and dairy is a major part of this productivity. (Although it's possible this productivity might change as California's water crisis escalates and as the political climate makes migration of agricultural laborers from Mexico and other parts of Latin America even more precarious.)

Our destination was a large auction yard a short distance east of Stockton. It was autumn, and the drive from the Bay Area passed through rolling brown hills and clusters of wind turbines. As we neared the auction yard, the landscape changed to flatland: farms and orchards extended to the horizon. As we zipped past dairy farm after dairy farm, the sea of black-and-white Holsteins blurred into a mass of hundreds, thousands, tens of thousands of bodies and lives brought into being as laborers for dairy production. I was reminded of a cross-country road trip my family had taken when I was a kid and the horror we felt driving through Kansas and Nebraska and seeing the steers who would soon be slaughtered for beef crammed into dirt and manure-filled enclosures stretching to the horizon, an endless cloud of flies buzzing above them.

The dairies we passed were a lot like Ansel Farm, but on a much larger scale. Like Ansel, there was no grass on any of the farms we saw—pens of dirt and manure stretched as far as we could see—and in them, many cows just stood there, in the sun, while others lay down in the dirt. Some ambitious cows stood just inside the fence on the side of the highway, reaching their heads through the fencing, stretching their tongues out to grab a blade or two of grass from the narrow patch of grass and weeds growing alongside the road.

Unlike at Ansel, there were far too many animals to count. At California's megadairies, it is not uncommon for a farm to house ten or fifteen thousand cows. The milking technologies necessary for this scale of production far exceed the twenty-cow-at-a-time milking capacity of Ansel. Rotary milking machines are increasingly common; these machines are circular, rotating platforms that turn as the cows are milked, moving cows on and off the platform in a constant stream.

The smell of manure was increasingly strong as we drove deeper into dairy country on rural highways. My mind traveled back to Homer Weston's words at Ansel Farm—*the smell of money*—and I remembered the tractor scraping liquefied manure into a pit, the truck that pumped it out of the pit and delivered it to the manure lagoons, and the problems of lagoon leakage into local waterways. That had been only a five hundred–cow dairy. The amount of manure that was managed at these larger farms felt unfathomable. The USDA calculates that the manure produced by two hundred cows used for dairy creates as much nitrogen as the sewage from five to ten thousand humans living in the United States.[1] It is, perhaps, difficult to imagine, then, the environmental impact of larger dairies—dairies like the hundred thousand–cow dairy being constructed in China (the largest dairy in the world to date).[2] Just to scale up from the USDA's statistics for the waste from two hundred cows, that would mean that a hundred thousand–cow dairy would produce nitrogen in amounts equivalent to that from the sewage produced by between 2.5 and 5 million humans.

California's water crisis, and its ongoing predisposition for drought, is exacerbated by dairy and meat production in the state. Animal agriculture is the most water-intensive sector of agriculture, which one might think would precipitate some conversation about cutting back on animal use for food production. Cows drink approximately five gallons of water for every one gallon of milk produced.[3] And in terms of overall water usage, it is estimated that it takes a thousand gallons of water for every gallon of milk produced.[4] However, instead of considering the water requirements of dairy production or other animal agriculture industries, other means of reducing water consumption have been promoted, such as individual homeowners limiting water usage, restaurants not serving water unless a customer orders it, and so on— all small gestures in comparison to the much more significant impacts of dairy and meat production.

Beyond water use and pollution, animal agriculture has a profoundly devastating effect on other dimensions of the environment. Methane emission from farms that house cows contributes in significant ways to air pollution and ozone depletion. Animal agriculture—particularly industrial agriculture—is implicated in the clear-cutting of global forests and soil quality. Land is cleared at an alarming rate not only for raising, housing, and slaughtering farmed animals but also for the production of feed crops: corn, soy, and other cereals. This widespread clearing of land for farming has deleterious effects on biodiversity and contributes to mass extinction. We are in what scientists are now calling the sixth extinction—the most catastrophic extinction event since that of the dinosaurs.[5] Farming animals at industrial and small scales is contributing in profound ways to this process.

Animal agriculture is also the leading cause of greenhouse gas emissions (and thus, contributes significantly to climate change), a detail that is conveniently ignored by many environmentalist organizations and individuals. Farming animals accounts for *18 percent* of global greenhouse gas emissions, according to the United Nations Food and Agriculture Organization—a larger percentage than all of transportation combined.[6] A documentary called *Cowspiracy* addresses the environmental impacts of animal agriculture and explores why environmentalists may be perpetually ignoring this problem.[7] California is a site not only where these environmental impacts can be seen but also where the urgency of this issue can be understood in the context of catastrophic climate change impacts.

As we drove along through California dairy country, the smell of manure stinging our nostrils, Tish and I marveled at the scale of production and the ensuing environmental impact, wondering about the connections among dairy production, the drought, and California's agricultural future. But this was forgotten for the time being when we spotted the sign for the auction yard and a new set of anxieties overtook us: the anxiety of fitting in at another auction and of anticipating the suffering we might see. When we pulled into the auction yard parking lot, I had the now-familiar anxiety of being out of place as I parked the little sedan rental car in among the rows of massive Ford pick-up trucks. We sat for a moment in the car and then, bracing ourselves for the stares and stress of the event, we headed into the auction. Sure enough, as we walked up to the pens containing the animals, the men standing around chatting stared at us. We smiled and politely said hello and then kept walking.

This auction had an expansive outdoor area, with a series of pens and chutes under a large pavilion roof, and then more longer-term holding areas out beyond the covered area. Above the covered pens, there was a long catwalk that stretched the length of the auction yard. We walked the length of catwalk, stopping intermittently to stare down into the pens to watch the animals waiting there. In the pens closest to the auction hall, there were groups of tiny calves, huddled together; some who were particularly frail lay on the ground, legs splayed out beneath them.

The rest of the pens housed groups of cows—mostly Holsteins— and a number of bulls. Many of the cows were emaciated, limping, with docked tails, and with udders that looked red and infected. Several had placenta cascading from their vaginas, having recently given birth. This sight, which is somewhat alarming to someone not accustomed to seeing it, occurs when the placenta is not fully delivered with the birth of a calf. Any parts of the placenta that remain inside the cow after the birthing process can cause infection. I felt for these cows, standing there in dirty pens, clearly having just recently given birth—their udders swelling with the milk meant for their calves. I wondered where their calves were and whether some of the tiny, spindly-legged calves in the calf area had been born at the auction to these cows or whether the cows had given birth at their farms and were then transported to auction.

In the other dairy auctions I had attended, it was unusual to see bulls. Full-grown bulls less commonly show up at auction yards (partly because there are not a large number of bulls who are raised up, intact, to adulthood), but this auction advertised online that they regularly sell bulls for breeding, which was one reason I wanted to visit. The actual bulls whose semen was advertised as such a hot commodity at the World Dairy Expo were entirely missing from the expo space, represented only by photoshopped catalog images showing off their reproductive prowess and body parts. But these were the real, embodied animals on whom the semen industry relies, waiting in the pens, impatiently snorting, stamping their feet, and, in pens that had enough room, pacing in tight circles.

About halfway along the catwalk, we looked down and saw a Holstein cow collapsed on the ground. Her back legs were splayed out behind her at an unnatural angle so that all her weight rested on her full udders, which were crushed beneath her, leaking blood and milk out onto the liquid manure–covered concrete ground. We stood and watched as she would gather her energy and try to rise, her back hooves scrambling on the slick cement, not gaining traction. The effort caused her to breathe heavily, and saliva foamed at her mouth. Her tail was docked, her back was covered with flies, and she lay there with no way to protect herself from the flies' biting. She had a yellow auction sticker with a barcode and the number 743 in large bold letters stuck to her side.

As we stood there on the catwalk watching her from above, a couple of the auction yard employees watched us watching her. They crossed the auction yard and entered the pen where she was lying on the ground. There were other cows, milling about in the pen around her, and the men reconfigured the fencing to herd the cows out into the adjacent pen. The nonambulatory cow remained, alone, on the floor of the pen.

The auction was beginning soon, and the men got called away, likely tasked with herding other waiting animals into the chutes that led into the auction hall. We left the catwalk, entered the auction hall, and found seats in the second row. We were the only people in the audience at that point, and the auctioneer sat behind the desk on his platform, preparing for the sale. The first animal was already in the auction ring, ready to be sold: a day-old Holstein calf with a barcode sticker #604.

The auction ring was clean: immaculate, really—because this was the first sale of the day, the wood shavings lining the floor of the pen had not yet been soiled by animals passing through. The tiny calf stood in the middle, looked around, and bleated. The calf's persistent cry was the only sound in the room, drowning out the distant noise of gates clanking and other animals bellowing from the pens outside. He was a newborn and his drying umbilical cord dangled from his belly as he stood there, slightly wobbly on his spindly legs.

The man whose job it was to herd the animals through the auction pen entered the ring with a paddle and leaned on the fence in front of the auctioneer's platform to chat with the auctioneer. The calf, noticing the man, approached him and gently nudged his leg. He stood no taller than the man's knee and his nose nuzzled gently but persistently at the side of the man's thigh. In one efficient motion, the man turned and smacked the calf in the face with the paddle he was holding and spoke (much too loudly for the fairly quiet room), "I'm not your mother!" The calf leapt back, cowering, and then ran away from the man, across the pen.

The man must have seen a look pass over our faces as we sat in the bleachers and watched, and he gave us an embarrassed smile and a nervous laugh. I felt compelled, for some reason, to smile back—a thin, forced smile. I was nervous that if I did not smile, I would be identified as someone who didn't belong. Smiling (or trying to) seemed to be an important gesture to demonstrate I was at ease in that place—not out of place. The silent awkward moment between us passed. The auction began and the calf sold—the first in a string of day-old calves who entered and left the auction pen on unsteady legs, fearful of the men who herded them.

As we sat there and watched sale after sale, I thought again about the way we learn certain ways of relating to animals that enable our use of them. When the calf approached the man with curiosity and requested attention, even though the man recognized this basic need for comfort (with his acknowledgment that the calf was seeking his parent), rather than providing a moment of care and intimacy, he rejected the calf's plea, smacking him in the face instead.

The first time I saw a day-old calf at auction was at a multispecies sale in Washington, where cows, pigs, goats, and sheep were sold all in the same auction. I had taken some students from one of my classes there as a field trip. As we sat in the bleachers together, I explained what was going on throughout the auction. A tiny creamy-reddish-brown animal stumbled into the ring and stood there silently. One of my students asked, "What's *that* animal?" At first, I didn't know—he looked like a small deer, but I knew that couldn't be right. And then I saw the drying umbilical cord dangling from his belly and I knew. Animal rights activists I knew had told me that day-old calves routinely showed up at auction and sold for very little, if they sold at all, but until then I hadn't seen one. And, to be honest, I wondered if it was one of those things that happens occasionally and is made to sound more routine than it is to emphasize the "horror" of dairy production. This small, skinny creature was a newborn Jersey calf, and my words caught in my throat as I told the students that this was a calf destined for veal production. He sold for $15. Since that auction, I've seen too many day-old calves to count; indeed, the selling of day-old calves for veal or beef production is certainly routine in the industry.

The way that calves are used in the dairy industry is first-and-foremost dependent on the way the industry determines sex at birth

(organized into a binary of male/female and reproductively viable/
not). I learned in my research that up to 5 percent of all calves in the
dairy industry die at birth.[8] Of those who survive, healthy female
calves are typically raised up into the industry as replacement heifers
for dairy herds. Females may be raised on the same farm where they
were born or they may be moved to another farm until they reach reproductive maturity. Farmers will often, in fact, sell females to other
dairy farms looking to increase their herd size. Young heifers sold as
"replacements" for dairy herds appear frequently at auctions, and of
all the animals I saw at auction, these heifers were in the best shape, as
they had not yet been subjected to the extractive and exhaustive rigors
of dairy production.

Male calves, in contrast, are of very little value to the industry and
are typically sold, or disposed of, at birth. Some male calves are castrated and raised as steers for beef. However, the majority of male
calves born in the dairy industry are raised in confinement and fed
milk replacers until they are sixteen to twenty-four weeks old, when
they are slaughtered for veal. When I first began my research on this
book, I was taken aback to find that there was such a clear connection between veal and dairy. And I was even more surprised that this
connection was so thoroughly unrecognized by the majority of consumers, many of whom claim that, by not consuming veal, they avoid
supporting its production. In fact, consumers of dairy support the veal
industry every time they buy a dairy product. It is important for the
continued success of the dairy industry that the dairy/veal connection
remains hidden.

Raising these calves as veal is a way to eke capital from the bodies
of male calves—bodies that would otherwise be considered "waste
products" in the industry. If a newborn male calf appears to be ill or
weak, he will likely die or be killed (often by firearm) at birth onsite at
the farm, as it is not economically viable to put time or resources into
his care. The bodies of these animals are either composted on the farm
or sent to rendering or, if these weaker calves seem to be slightly more
robust, they will be kept alive for just a few weeks and slaughtered for
cheap meat, called bob veal. Bob veal is a less expensive alternative to

special-fed veal and is the flesh of extremely young calves. Approximately 15 percent of calves raised for veal are slaughtered for bob veal and do not live past three weeks of age.[9]

It is unusual for dairy farms themselves to be involved in the raising and slaughtering of male calves for veal. Instead, they will generally sell day-old calves at auction to buyers interested in raising them, or they will contact a buyer to purchase the calves directly from the farm. The dairy industry fosters the invisibility of male calves and conceals the link between dairy and veal production. Because the ethical dimensions of veal production have received significant public attention since the 1980s in the United States, the dairy industry attempts to publicly distance itself from the tarnished veal industry. The mistreatment of calves in the veal industry was brought to the US public eye in the 1980s and 1990s, when it was revealed that calves raised for veal were kept under low light, chained in tiny crates to immobilize and isolate them from one another in order to keep their flesh pale and tender and free of infectious disease. Despite public disfavor, this is still primarily how calves raised for veal are treated—isolated and confined in individual crates or hutches.

On more than one occasion during my research, I encountered evidence of the dairy industry's attempts to distance itself from veal. At Ansel Farm, I asked Homer Weston what happened to the male calves born at the farm. He replied that a buyer came to the farm to pick up the calves, but that he didn't know—or ask—how these buyers planned to use the calves. However, when I asked if they might be used for veal, he quickly replied that their farm had nothing to do with veal and didn't want anything to do with it. At the World Dairy Expo, too, I inquired about raising calves for veal, and industry representatives were hesitant to talk about the dairy/veal connection. This distancing from the veal industry is important for the maintenance of a positive image of dairy production. For, of course, there is something wholesome and empathy evoking about calves: maybe the fact that they're so young, that they're fragile and cute, and that they're vulnerable prompts more acknowledgment of their need for and deservingness of care. And there's maybe something deeper at work, too:

maybe those humans who bonded to their parents, who once cared for them as infants, can relate to the violence of severing this intimate bond.

Through my research, I came to expect a level of silence and secrecy around the practices of dairy and veal production in large-scale, industrial contexts. These industries are fearful of negative publicity and the release of unsavory information that may harm their business. What still surprises me, though, even after doing this research, is the attitude of small-scale, local producers of dairy and meat: the simultaneous feigned transparency I encountered and the unwillingness for most farms to show me their production practices. But this is also paired with a pervasive trend of naturalizing the use of animals and pushing back against the growing knowledge that raising animals for food is an inherently violent process, including, as already alluded, the case of veal production.

In the Pacific Northwest and elsewhere, in the local food movement, where consumer connection to food is touted as essential for ethical eating, there is a growing conversation about so-called ethical veal. This is, of course, part of the larger trend among meat eaters who believe that an ethical method of killing animals for food is possible through small-scale farming. In 2012, a story ran in the magazine *Edible Seattle* that featured veal producers who buy and raise day-old calves from local dairies. The article's author, Amy Pennington, argues that this "ethical veal closes the loop at local dairies."[10] The focus on local veal production draws attention to the connection between dairy and veal in a way that is unusual—and in a way that may cause consumers to engage more directly with the fuller network of industries tied to dairy production. But this move to eliminate the taboo associated with eating veal works to further naturalize the use of animals in the dairy and veal industries. The article's title—"Consider the Calf"—represents an excellent example of doublethink, expertly executed. Pennington asks us to see the calf's role in veal production as a necessary part of dairy farming. So, we're meant to *think* we're seeing and considering the calf. In fact, what we're seeing is the calf as a site of consumption and capital accumulation: an untapped source of income for the farmer and a tasty delicacy for the consumer. As soon as we see the calf in this way, we are immediately required to forget that all calves (indeed, all the animals in meat and dairy production) are unique, emotional subjects of their own lives.[11] The calf himself, of course, is not considered at all. What is considered is what he can provide for us (the consumer, the farmer, the dominant human).

Sea Breeze Farm is a tiny farm on Vashon Island, Washington, selling at Seattle area farmers' markets. This, in fact, was one of the farms that first encouraged me to visit and then repeatedly put me off when I contacted them to schedule a visit, eventually denying me even the most basic tour of their farm, even though their website, for several years, encouraged the reader to check back soon for tour dates. The owner, George Page, works hard to naturalize the process of farming animals—to make it seem as though he is returning to nature, which appears to mean going back to "the way things have always been" and which, to those accustomed to it, would, of course, seem to be the natural way of doing it. This process of naturalization is evident in a news article about Page: "In his view, large-scale modern farming has segmented animal husbandry, breaking up what was once a whole and healthy ecosystem. Byproducts—such as male calves on a dairy farm—then become problems rather than just pieces of the puzzle. At Sea Breeze, he wants every piece to play its part."[12]

Indeed, industrial production and its segmentation have attracted a great deal of attention to the politics of food and agriculture and pushed consumers to rethink their consumption habits. In the case of dairy and veal, it is precisely the segmentation of these industries that has both revealed and concealed the violence of the process of raising and killing animals for food. Industrial production and its division reveal the violence of the system that becomes visible to consumers when certain aspects of these industries are exposed (e.g., the recently mentioned exposés of veal production in the 1980s). But the segmentation of the industry also enables the concealment of other kinds of violence. For instance, the division between the dairy and veal industries works to the dairy industry's benefit because dairy can continue to sell well when consumers boycott veal for ethical reasons.

Disconnecting from the reality of the calf's life (crated and perhaps offending our sense of how animals should be treated) makes it possible to imagine that dairy production does not involve any of these more unsavory practices. This division between the two industries, whereby male calves become "problems," in effect emphasizes the violence of human-animal relations in this context.

In the case of Sea Breeze, where male calves become "just pieces of the puzzle," what is involved in humans raising and killing animals becomes naturalized by the suggestion that this is a return to a "whole and healthy ecosystem." Rather than asking consumers concerned with the ethical, political, and environmental dimensions of animal use in the food industry to *move forward* to radically rethink the system itself, consumers of niche, local animal products are reassured that eating veal is actually a way to *go back*, to get closer to nature—to close the natural loop of dairy-meat consumption. This imagined alternative of "ethical meat," however, is still situated in a hierarchy of dominance and subjugation, whereby humans exercise power and control over animal lives, albeit in a more aesthetically palatable package.

The phrase "humanely raised and slaughtered" works aesthetically and discursively to obscure the violence of meat, dairy, and egg production in any form. Further, this trend in consumer-food connection results in desensitizing consumers to how using animals for food is violent, perhaps even more so than in the mainstream industry. The mainstream meat, dairy, and egg industries rely on a lack of consumer knowledge about the role of animals in these industries and paint generic images of the "family farm" to reassure consumers at the site of consumption—the grocery store. The niche, local suppliers of "ethical" or "humane" meat, dairy, and eggs, conversely, operate under the guise of full disclosure, encouraging consumers to acknowledge the process of killing animals for food, while at the same time necessarily denying the system of human exceptionalism that this act of killing requires.

There is something noteworthy and emotionally affecting about the calf raised for veal. There is something in the reality of this newborn animal being housed, isolated from his parent(s) and others of his kind for a short life before he is slaughtered, that offends many people's sensibilities and appeals to their emotional responses. Rather than deny this emotional response, I would suggest that bringing the calf raised for veal—as well as the bull used for semen, the cow for dairy, the chicken for eggs and meat, the turkey for Thanksgiving dinner, and the pig for bacon and pork—into our sphere of empathy is a deeply political act of refusing to engage in everyday practices that dominate and violate other species for our own gustatory pleasure. The life of the calf is an entry point, then, into expanding our circle of whom we care for and about to more distant others with whom we are intimately intertwined every time we choose what we eat.

Prior to arriving at auction, the day-old calf mentioned earlier, with ear tag #604, would have been taken away from the cow who gave birth to him just hours after; in fact, it is the recommendation of the Bovine Alliance on Management and Nutrition that calves and cows should be separated within the first hour after birth.[13] Homer Weston at Ansel Farm explained that they remove all calves so quickly after birth because the animals bond more closely the longer they are allowed to be together. As it is, Homer explained, cows will bellow for up to two weeks for their calves after they are separated. Homer acknowledged the trauma experienced by animals after separation and the deeply emotional nature of bovine relationships.

When calves are born, their blood does not contain the antibodies needed for a healthy, functioning immune system; calves get these antibodies from drinking colostrum, the first milk produced by the cow after birthing. Colostrum has higher concentrations of protein, fat, vitamins, and minerals than regular milk and gives the calf essential nutrition immediately after birth.[14] One would think that the easiest way for calves to obtain colostrum would be to let them feed from the cow immediately after birth. However, concerns about bacterial infection, transmission of disease, contaminated colostrum, and the intensification of the cow-calf bond motivate the logic behind preventing calves from feeding directly from the cow.

Colostrum replacers were prominent at the World Dairy Expo. These products are made from dried bovine colostrum or serum and are usually effective (though not ideal) in providing the calf with the necessary nutrition and antibodies. When calves are transitioned off colostrum, they are generally fed a milk replacer until the time when they are weaned onto a dry, calf starter feed. Milk replacers contain a range of ingredients: animal fat and vegetable oil, animal plasma, casein, dried skimmed milk and whey, lecithin, and vitamin and mineral supplements.[15] Additionally, milk replacers may contain mixtures of animal and plant protein products (including fish, meat and bone meal, dried blood, soy, cottonseed, and brewers' yeast), products that are extracted during the rendering process I explained earlier in the book. These milk replacers are mixed and fed to calves in bottles or buckets until weaning. Often, it is difficult to get calves to drink the milk replacers, and so companies like the ones I encountered at the World Dairy Expo have emerged that produce flavor additives for colostrum and milk replacers to encourage calves to drink the formula.

More than 60 percent of dairy producers in the United States wean calves off the milk replacers at eight weeks of age or more, while fewer than 30 percent wean calves at six weeks.[16] Based on a calculation of total feed per calf for the first eight weeks of life, feeding a calf a combination of milk replacer and calf starter feed and weaning at eight weeks costs approximately $15 more per animal than weaning at five weeks.[17] It is economically advantageous, then, to wean calves as early as possible, while still maintaining their health. Colostrum and milk replacers are one example of how the division of the industry results in biological processes among animals being co-opted and commodified by humans, creating replacement products for this natural feeding process.[18] In fact, claims that milk replacers perform better than cows' milk is one of the many examples of a techno-fix in the industry, where pressures put on the cow's body are mitigated or obscured by technological innovations designed to replace or augment the biological functions of the cow's body.

The veal crates that made US news in the 1980s and 1990s—crude wooden crates just big enough to fit a calf—are still used, but the dairy and veal industries are now publicly promoting the more innocuously termed *hutch* (and indeed, one company makes enclosures for calves that they advertise as Comfy Calf Suites).[19] Some farmers have transitioned from wooden crates to plastic crates or hutches because they are easier to clean, are lightweight, and disassemble easily. Calf hutches look like large plastic domed doghouses and are marketed for raising calves for dairy herds but can be used in raising calves for veal. Calves, like other nonhuman and human adolescents, are playful, and seek the contact and companionship of others. In the case of the traditional wooden crate, they have very little mobility. With the hutch, they are typically chained to the front so that they can move inside and outside the hutch. However, because of their playful nature, a representative for Calf-Tel at the World Dairy Expo mentioned that calves regularly hang themselves on the chains while trying to play with calves housed in neighboring hutches. To avoid this unfortunate outcome, she recommended small fence enclosures—instead of chains—for containing calves in the hutches. Whether calves are housed in rows of crates, hutches, or small pens, though they can see the other calves near them, they are restricted from any contact with one another to prevent, as mentioned earlier, the spread of disease and contaminants among calves.

Female calves, too, are raised in isolation in hutches until they are old enough to join the dairy herd. This isolation goes against the way bovine animals prefer to live. Cows and other bovine species are herd animals and will remain together as a herd for their lifetime if allowed. They form cliques, have close emotional bonds, and form lifelong friendships. Raising calves in hutches restricts their species-specific behavior preferences and isolates them from much-needed companionship. On some farms, calves will be transitioned to group housing in a larger hutch or pen with other calves once the perceived threat of disease transmission has passed. This is the case especially when females are being raised as replacement heifers and socialization is necessary to acclimate them to living in herds with animals after being raised for a period in seclusion.

The calf with ear tag #604, nuzzling the man for comfort in the auction ring, was demonstrating this predisposition for sociality. The calves in hutches who hang themselves on the chains trying to get to their neighbors are demonstrating a need for play and a desire to so-cialize. The auction yard, and the segmentation of the dairy and veal industries more broadly, have the effect of alienating animals from one another.

* * *

After the auction was over, we went back to check on the collapsed cow in the holding pen. Again, we stood on the catwalk above her and, this time, noticed that someone had bound her back legs together with a blue cord while the auction was going on, in an effort to help her rise. But she was still much too weak, and the cord only enabled her to turn on her side with both legs together so that they weren't splayed out behind her. If she couldn't get up, she would be shot with a firearm at the end of the day and her body would be transported to the nearest rendering facility.

As we watched, we heard a commotion behind us. A Holstein bull had broken away from the others as they were being loaded into the transport truck at the back of the auction yard. We had seen him sell in the auction, along with three other bulls for breeding. Spooked by the noises and the men with electric prods herding them, he bolted back down the corridor, trying to escape. The men scattered, frantically try-ing to get him to veer off, yelling and slamming a gate closed just as he tried to pass through. Trapped, the bull trotted back and forth, looking for another way out. He was young, healthy, and strong, his coat shiny and thick, and his muscles visible as he trotted back and forth in the enclosed chute.

Though he was much stronger than the men attempting to load him into the transport truck, the layout of the auction yard was de-signed to subdue the powerful strength of thousand-pound animals. The men were breathless from the chase, and they were angry at the bull for his escape attempt, cursing loudly. One of the men held an electric prod and jabbed the bull repeatedly with it. Leaping forward with each shock, the bull was forced up the ramp into one of the truck's compartments. Once the bull was in the compartment, the man—

unnecessarily so—jabbed him several times with the electric prod through the holes in the trailer and the bull lunged against the side of the trailer walls trying to get away from the prod. Finally, the men closed the trailer and the truck slowly pulled away from the auction yard out onto the open road on its way to the breeding farm where the bulls would be kept for semen production.

A moment after the truck pulled away, the men who had been herding the bulls turned their gaze up to the catwalk where we stood. Clearly frustrated by our focused attention, one of the men yelled at us, gruffly, "Hey! Can we help you girls with something?"

"No, thanks!" I called back, "We're just watching."

"Well, there's nothing to see here," the man replied, "so unless we can help you, the auction's over . . ." His voice trailed off, but we had gotten the gist of what he was saying. He wanted us to leave, and, shortly after, we did leave, exhausted, and thinking about the cow who couldn't get up and who would die at the end of the day.

10

ON KNOWING
AND RESPONDING

When I was growing up, some of our closest family friends were a family of very strict vegetarians and environmentalists. Their son and daughter were roughly the same age as my sister and me. We adored them (and still do) and spent long hours playing in the woods outside of Pittsburgh, going on nature walks where we would find fossils and splash in streams, looking for salamanders and frogs, and build forts out of sticks and leaves.

During these years, we went regularly to the East End Food Co-op in Pittsburgh where our parents shopped, and when we went along with them, their mom, who was like another parent to us, would buy us a treat—always carob, never chocolate. I remember thinking, "Oh geez, carob again? What's wrong with a little chocolate?" but in the end I would eat it, happy to have something sweet no matter how "weird" I thought it was. In addition to being vegetarian (they did eat dairy and eggs), they also tried hard to avoid added sugar, caffeine (in the form of coffee, chocolate, and tea), processed foods, and other food additives—conscious not only of the health impacts of those foods but also of their human labor and environmental costs.

They sought out organic food before "organic" was a thing. Their use of canvas bags for grocery shopping (something my mom also did in our home) stood out as something unusual back then. And they traveled with these canvas bags full of groceries so they would have food they could eat wherever they went. The food choices they made were carefully thought out in terms of how they would affect the environment, the people producing their food, and their own health.

It wasn't that our family was unconcerned with these things—we had an abundant garden that produced fruits and vegetables to feed us through the summer and into the fall; we composted; we ate foods (like natural peanut butter and dense homemade whole wheat bread) that my classmates at school thought were weird; and we gave out boxes of raisins instead of candy at Halloween (yes, we were *that* house). My parents would rinse their paper coffee filters after they made coffee and dry them in the dish rack to save paper. In fact, I think one box of fifty coffee filters lasted years in our house, as did a single roll of unbleached brown paper towels. The paper towels were mostly reserved for cleaning up cat vomit, and we knew never to use them except for that. If we had guests who tore off some paper towels after washing their hands at the kitchen sink (instead of using the cloth dish towels), my dad would swoop in, grab the wet paper towels from the guest's hands, scowling and muttering, and smooth them out to dry in order to be used again by the next "environmental criminal" (his words) who darkened our door.

So, it's not to say that our family didn't engage in our own set of environmentally minded practices that probably seemed unusual to outsiders, but I remember thinking that our family friends' food choices were way more extreme. I couldn't understand why they couldn't just eat meat like the rest of us. *What was the big deal about eating a little meat,* I wondered, *weren't they missing out?*

It never crossed my mind to wonder, instead, why we ate animals. It actually rarely crossed my mind, in any significant way, that what we were eating had, not long ago, been a living, breathing animal. This, while I claimed to be a dedicated animal lover. For instance, as a child, I was horrified by the number of potato bugs (at least that's what we called them; they're more commonly known as roly polies or pill bugs or, if you want to be scientific, *Armadillidiidae*) who were squashed while trying to cross the sidewalk leading up to the back door of our house. And so I made "POTATO BUG CROSSING" signs for both sides of the sidewalk and I would remind people with great urgency, as we approached the path, to please be careful not to step on any. I remember one time one of my best friends intentionally jumped on—and smashed flat—a potato bug that I pointed out to him to avoid stepping on. I was so furious I didn't speak to him for a week.

At the same time that I cared so deeply for the potato bugs, and the squirrels we named Pouchy and Alice who would feed in the backyard, and the ducks at the park, and the cats and other critters we kept as pets, I still ate animals without thinking twice. I made a half-hearted attempt when I was maybe twelve or thirteen to declare that I was now a vegetarian, but I think that was more a mark of rebellion than it was an informed ethical or political decision. And it was short-lived. I quickly went back to eating animals, not questioning it again until much later.

Now, looking back, I marvel at the way our family friends maintained their ethical and political commitments in the 1980s, decades before vegetarianism and veganism were commonplace. There are parts of the United States where I've traveled in the last ten years and been hard-pressed to find a vegetarian or vegan option, but generally the word *vegetarian* has become familiar to most people—even if it is still often a target of disdain. Things have changed a lot in recent decades, and when I talk to people who have been vegan for twenty or thirty years, they talk about how much easier it is to be vegan now compared to previous decades. The availability of plant-based foods is widespread, and the stigma around being vegan is beginning to wane as more people adopt a plant-based diet and as the health benefits and the environmental sensibility of forgoing animal-derived foods becomes more widely acknowledged.

After I started researching the plight of animals in the meat and dairy industries, my mom and her partner, Jim, drastically cut back on their consumption of animal products. My mom talks about their transition away from animal-based foods as a kind of convergence in thinking. She has always been environmentally minded and very health focused. For both of those reasons, well into her sixties, she walks four miles to work and back every day, using her already-fuel-efficient car as infrequently as possible. When she learned that animal agriculture contributes far more significantly to climate change than even transportation, this caused her to pause. And then when she began learning about the ethical dimensions of raising animals for food—in other words, when she faced the reality of the way animals live and die in the food industry—she says that something just clicked for her and she didn't want to participate in perpetuating that violence anymore. This just reinforced the ideas she already had about nonviolence, environmental consciousness, and compassion.

Jim is a meat-and-potatoes kind of guy—always has been. He is also a type 2 diabetic (now in remission). For him, the most compelling information he encountered was the recent research on the health benefits of a plant-based diet. As it turns out, the consumption of animal products is implicated in many of the major killers—diet-related diseases like heart disease, diabetes, and cancer. *The China Study*, written by Thomas Campbell and T. Colin Campbell, was one of the largest epidemiological studies in history and the first comprehensive study of the relationship between diet and cancer growth, showing that regions in China where animal consumption was uncommon had dramatically lower instances of cancer. Since then, other medical professionals and researchers have conducted studies on the impacts of a plant-based diet to prevent, treat, and reverse diseases like heart disease and diabetes, among other common ailments like high blood pressure and elevated cholesterol. Dr. Joel Fuhrman, for instance, author of *Eat to Live* along with many other books, takes a "nutritarian" approach to disease prevention and treatment, recommending a nutrient-dense, whole foods, plant-based diet for health and longevity. This was the book that really spoke to Jim, and in six months of eating this way, Jim was able to have his diabetes, blood pressure, and cholesterol medications dramatically reduced by his doctor. Of course, I'm not a medical doctor and so cannot comment on the medical or scientific effects of this way of eating, but there is a growing community of medical professionals interested in studying and working on these approaches to health and disease.

I think everyone comes to information like what I've shared in this book differently, based on our personal histories, the way we were raised, the values we each hold most dear, and, quite frankly, our personalities. When I first started learning about the plight of animals in the food system, I couldn't contain my grief. I was devastated as I read

and watched everything I could find about animal agriculture. There
are images I saw years ago that will be seared in my mind for the rest of
my life. There are certain things that, when you learn them, shift your
view of the world completely. And you have no choice but to respond.
Forgetting becomes an impossibility.

This was one of those things for me. When I learned about the way
fish, chickens, turkeys, ducks, pigs, goats, sheep, and cows suffer for
food production in the United States and globally, I couldn't believe
that life was going on as usual around me. Quickly forgetting my own
very recent ignorance and apathy about these things, I wanted to shake
people around me and say, "What is *wrong* with you? Don't you know
what's happening here? Don't you *care*?"

The grocery store became a site of mourning: the innocuous refrig-
erators filled with milk, yogurt, cheese, butter, and eggs; the freezers
of ice cream; the cases of meat, neatly packaged and priced—these
suddenly became, to me, the products of immeasurable violence.
Philosopher and critical animal studies scholar James Stanescu talks
about the sense of grief over animals suffering provoked by the experience of grocery shopping and its illegibility to most people because
of the way violence against animals is made so routine and normalized.[1]

There were many times when I went grocery shopping, and, passing through the meat or dairy sections, I would freeze, my eyes well-
ing up with tears, and I would hurry out of the store, abandoning my
grocery cart, buying nothing. This can feel immensely isolating as it's
a response that most people think of as silly, overly sensitive, or down-
right unhinged.

And this is maybe one of the more difficult things about becoming
conscious of the scale of suffering and violence inherent in food production and, for that matter, in many modes of commodity production
globally. The isolation of mourning for the routinely unmourned is
significant. Learning about sweatshops, child labor, human trafficking,
global poverty, environmental disasters like oil spills or clear-cutting,
climate change, mass extinction, war, agricultural labor—all conse-

quences of global (and particularly Western) consumerism, or capitalism more broadly—can shatter one's faith in humanity.

These are grief-inducing global problems, just as animal agriculture
is, and attention to the grief one may feel is an important part of showing their political dimensions. Judith Butler's books *Precarious Life* and
Frames of War outline the way grieving for the deaths of non-US casu-
alties of the War on Terror makes a profound political statement that
Iraqi and Afghani lives and deaths matter and have meaning. Butler's
and Stanescu's work has helped me to frame how I think about the
grief I felt for animals in the dairy industry. In some ways, learning
about animals in the food system (or any other global problem of epic
proportions) means entering a process of perpetual mourning.

I often hear people say things like, "Well, yeah, obviously that's
super sad, but I can't care about *everything*! I choose the issues I care
about and try to change them." In some ways this is a pragmatic and
sensible response: each person does have a limited amount of time,
energy, and personal resources and each person feels particular affin-
ities for certain issues based on their life experiences and how those
issues were introduced to them. But this siloing of issues—the idea
that significant progress can be made by working on just one—ignores
the very real ways in which forms of injustice are interconnected and
mutually reinforced.

For example, working against modes of systematic killing—like the
way human lives become collateral damage in war—may very well be
inhibited by the fact that other forms of systematic killing (like animal agriculture) are so widely accepted and normalized. If it is already
commonplace to say that certain lives (in this case, animals) just inherently matter less, then it becomes possible to make the stretch and
say that some *human* lives matter less, and then it becomes possible to
call the mass loss of human life something as innocuous as collateral
damage. And so it doesn't make sense, for instance, for an antiwar activist (who is otherwise committed to nonviolence and who is working diligently against the way killing is made routine) to ignore their
daily participation in the systematic killing of billions of nonhuman

animals. These and other forms of violence are predicated on categories of the animal, the subhuman, the less-than-human that render
lives killable—without interrogating and dismantling the material and
discursive work these categories do, there are limits to what progress
can be made for human or nonhuman animals.[2]

There's a tendency, too, in some threads of the environmentalist
movement, to ignore the impacts of animal agriculture and to focus
instead on personal, individualized conservation practices, like saving
water in the home through shorter showers, changing lightbulbs to
energy saving ones, or cutting down on the use of plastic bags at the
grocery store. While these are certainly important steps to take, the
impacts of a shorter shower pale in comparison to the water used in
raising animals for food (not to mention the degradation caused by
animal agriculture to air and soil quality). But environmental organizations shy away from animal agriculture, perhaps because it is too
political and seemingly asks too much of people, as *Cowspiracy* reveals.

Within the animal rights movement, as within other social justice
movements, there remain persistent problems of racism, privilege, cultural unawareness, and a "post-racial" attitude. Amie Breeze Harper
has written in different contexts on racism and whiteness in the animal
rights movement and calls out this post-racial attitude as one that ignores (and thus perpetuates) systemic racism.[3] This emerges, in part,
as a result of a tendency to prioritize animal concerns over all others,
enacted, for instance, in the vilification of workers in animal-use industries or in the false assumption that veganism is a predominantly
white, middle-class phenomenon. Working for animal justice issues
will continue to reproduce these uneven relationships of power, racial-
ization, and harm unless racial justice, gender justice, and other forms
of human struggles over equity become more central commitments
in the movement. Seeing the ways in which the exploitation of animals is intertwined with other forms of marginalization and oppression is integral to understanding the varied ways in which violence is
enacted through capitalism. Far from detracting from nonhuman animal issues, a consideration of how these forms of injustice mutually

reinforce each other can deepen meaningful work for a less violent and oppressive way of living and being in relation with others.

When I teach classes on animals in the food system and we talk about these intertwined forms of violence and their relationship to capitalism, one of the first things my students want to know is, "What can we *do*?" We are a species who likes to fix things, to innovate, to problem solve. It's hard to sit with a problem and think about its shape and form without thinking about how to fix it. Because the issue of animal agriculture is one entangled so much with consumption, students' first thoughts are often toward radically changing their consumption practices by partially or completely eliminating animal-based products from their lives. We talk a lot about veganism and plant-based diets. A number of students leave my classes each term determined to cut animal consumption (in the realms of food, fashion, and other areas) out of their lives.

Veganism can, indeed, be a powerful political statement about how a person values the lives of others and it acts as a marker that another kind of relationship with animals is possible. Like the sanctuary model that radically reimagines what it means to live in community with other species, veganism is a way to embody an alternative ethic in daily life about animals, food, the environment, health, and what kind of world we want to inhabit. It's not uncommon to believe that veganism is an extreme response—just like I thought our family friends' vegetarianism was extreme in the 1980s—but what is actually extreme is the fact that our insistence on consuming animal products is facilitating the escalation of climate change at devastating rates, the skyrocketing costs of health care due to diet-related disease, and the intensification of industrialized capitalism, which causes mass suffering for animals and racialized human laborers in spaces of food production.

And yet, the simple act of buying differently—the increasingly popular adage of "voting with your dollar"—does not fundamentally challenge the exploitative nature of capitalism. Plant-based food, for instance, does not ensure the fair treatment of the farmworkers producing fruits and vegetables or the distant laborers living in conditions of slavery or otherwise harmful conditions to produce nondairy chocolate, coffee, sugar, or bananas. And so, while shifting consumption choices away from eating animals is an important and productive step (especially as a means of responding to the widespread suffering of animals and the environmental crises of animal agriculture), much more is needed to make change on a systemic, less individualized level.

Adopting a vegan ethic is just one mode of making change, but it may not even be a viable action for those living in situations or places without good access to affordable, healthy food, and, as Kymlicka and Donaldson point out, "such choices may be very costly in social or material terms for certain groups."[4] Unfortunately, it remains more affordable and accessible to buy a fast food burger than broccoli, a symptom of government subsidies for meat, dairy, and highly processed foods composed mostly of corn. Projects to improve food access—and particularly food production outside of the capitalist economy—are vital and take many forms. From the creation of public community gardens or places like the Beacon Food Forest in Seattle (designed to be a self-sustaining forest of food for any and all who want to harvest it) to the redirection of the 40 percent of edible food heading to landfills in the United States into local food distribution networks (as does Food Shift, an organization in the San Francisco Bay Area), alternative food production and distribution practices are modes of shifting access to healthy foods.[5]

Changes in local and federal policy are obviously another way to make some kinds of change, and petitioning local and regional political representatives to make food access a part of the political agenda may work to change some of these problems at a systemic level. Changing thinking about health and disease prevention is another site of possibility. The Fruit and Vegetable Prescription Program at Harlem Hospital Center in New York is an exciting move in this direction. It is a program that prescribes fruits and vegetables as a health improving strategy, supplying patients with Health Bucks that offset the costs of produce (providing patients with twice the buying power of food stamps at local farmers' markets).[6] The organization Food Not Bombs is an international anarchist grassroots response to hunger, distributing free (usually plant-based) meals to homeless and food insecure people around the world.[7] Food Empowerment Project is a vegan food justice organization working simultaneously for vegan, environmentally sound, and socially just food production practices and policies in the United States and beyond, and working on farmworkers' rights, on healthy food access, and to end slavery in the chocolate industry (among other exciting initiatives).[8] There are sanctuaries—like Animal Place and Pigs Peace Sanctuary and Farm Sanctuary—working to educate visitors and give a life worth living to the animals who reside there, and there are places like VINE Sanctuary that are working to further reimagine and disrupt the power relations at work between humans and other species in sanctuary settings. And then there are people like Harold Brown, and his organization Farm Kind, helping farmers transition away from animal agriculture into other forms of farming economies, showing concern for the livelihood and survival of US farmers and farming communities.

There are all these efforts and more, working within and outside of the already-existing system to try to transform the dominant ways of relating to other species. There are local, regional, national, and international movements for animals, for food justice and food sovereignty, for decolonizing the diet, and for more multispecies and environmentally attuned ways of being in the world. These are efforts that already exist and with which it is possible to be involved by engaging with their work.

But studying the lives of animals in the dairy industry has prompted more fundamental questions for me about violence, commodification, care, and knowledge production. How do practices that visit violence on a body and life become normalized and routine—so much so that the violence does not seem like violence? In the case of the dairy industry, violence is normalized by a constellation of economic, political, and social frameworks. The economic logics that render the cow a commodity obscure, through commitments to efficiency and capital accumulation, the violence at the root of practices that are integral to the commodification process that harm the cow, calf, steer, and bull (artificial insemination, impregnation, separation, intensive milking, slaughter). Conceptualizing a life as a commodity limits the way of knowing that life: as a commodity, that life is understood in terms of what and how efficiently they can produce. They become a form of living capital and their care (no matter how caringly delivered, as in the case of Homer Weston's kindness and clear love for his cows) necessarily must be oriented around facilitating efficient commodity production. Legal framings of farmed animal species as property and as minimally protected under welfare regulations make many of these practices not only legal but also normalized and socially accepted through their legality. Advertising and industry discourses normalize industry practices through humor and narratives about the necessity of dairy consumption. Social mechanisms of education and tradition naturalize farming animals and connect practices of animal agriculture to nostalgic histories of wholesome intergenerational family farming (a process that not only makes violence against the animal in farming seem natural but also elides the violence of settler-colonialism that underwrites ranching and farming in the United States).

Within this framework of understanding the violence of commodification, how can new knowledges and forms of care with other species be imagined? In researching Animal Place and Farm Sanctuary, working with Pigs Peace over the years, and recently visiting VINE Sanctuary, I have been moved by the alternative ways of knowing and caring for and about farmed animal species that can occur in sanctuary settings. Routing animals out of the commodity circuit (and thus, out of being conceptualized as commodities) reorients how caring relationships are allowed to manifest and how knowledge about other species and singular animals is conceived. But even within the sanctuary context, Donaldson and Kymlicka ask questions about how sanctuary spaces might be imagined to be more intentional communities in which interspecies relationships can flourish even more purposefully and creatively.[9]

Building on the possibilities that emerge from thinking about caring relationships and knowledge production differently in encounters with nonhuman animals, what aren't we thinking of? What does it mean to care for nonhuman animals in ways that are not oriented around human interests? How might care be reconceptualized and re-evaluated in careful and ethically attuned ways? What can be learned about attending to the intimate and embodied lifeworlds of other animals through careful considerations of the spaces and communities in which they live? At root in these questions must be a fundamental commitment to denaturalizing violence against animals and decommodifying their lives and bodies. My hope is that the singular animals who are most affected by this violence and commodification—the cow with ear tag #1389, the day-old calf seeking comfort in the auction ring, the bull shocked with electric prods, the cow and the calf sold separately at the auction, the heifer waiting to give birth at the farm, and Sadie—can be a guide. My hope is that it is possible to learn from them, to let their stories be instructive as to how human-animal relations might be radically reimagined. My hope is that they prompt us to respond.

ACKNOWLEDGMENTS

Writing this book has been a truly collaborative experience—from sitting down to write in the company of my human and nonhuman animal comrades to the countless conversations, caring companionship, and inspiration from many others who have taught me so much. When I set out on this book project, I wanted to write a book that would appeal to a general audience; I wanted to tell the story—my story, and a shared story—of studying the dairy industry. I owe enormous gratitude to University of Chicago Press for publishing this work and to Christie Henry, in particular, for her editorial guidance and support. Christie's belief in this story being told and her calm, collected advice, made its publication—and, more fundamentally, its *writing*—possible. The publication process would also have been impossible without the many different hands this passed through in the editing process (Doug Mitchell, Kyle Adam Wagner, and Yvonne Zipter) and promotion (Tyler McGaughey).

This is a much stronger piece of work as a result of the care and time spent by several reviewers. Timothy Pachirat offered invaluable and substantive feedback on this manuscript at various stages, pushing this to be a much better book than it would have been otherwise. Yamini Narayanan's enthusiasm, comments, and commitment to this work have been incredibly meaningful. And the productive criticism from an anonymous reviewer pushed me to make this manuscript stronger. Although not a formal reviewer for the book, I am also enormously grateful for Logan O'Laughlin's detailed and careful reading of this manuscript and attention to ways of deepening its impact.

Special thanks to Jo-Anne McArthur and the We Animals project for allowing me to use her photograph for the cover. Huge thanks also to Anne Sullivan for her work on publicity and promotion.

I am indebted to the anonymized people who were interviewed for this work, at the sanctuaries and at Ansel Farm, for sharing their wisdom and knowledge and for making this project possible. I am also especially grateful to Marji Beach and Gene Baur for their enthusiasm and generosity of time in talking about this work and reading earlier versions of some sections. Judy Woods not only has been a clear source of knowledge and insight but has also become a wonderful friend—a member of my chosen family.

I owe enormous gratitude to Lori Gruen and Wesleyan Animal Studies for giving me the time and space to work through the revisions for this book, and I am especially grateful to Lori for her thoughtful conversations about this work and the warmth of her friendship through its completion.

I have been wonderfully lucky to work with and learn from such an extraordinary group of brilliant scholars, activists, and friends over the years: Michael Brown, María Elena García, Victoria Lawson, Rosemary-Claire Collard, Lucy Jarosz, Yolanda Valencia, Anika Lehde, Magie Ramírez, Amy Piedalue, Juawana Grant, pattrice jones, Carol J. Adams, lauren Ornelas, Annie Dwyer, Larry Knopp, Brandon Derman, David Giles, Tony Hatch, Krithika Srinivasan, Mónica Farías, and Elaine Jessie. They have each taught me so much about radical forms of care, love, and a commitment to what is just.

I have worked with some amazing students who have since become good friends and who have thoroughly shaped and inspired my thinking. From them, I continue to learn all the time: Sarah Olson, Dylan Forest, Meghan Jones, Katie Bartel, Mei Horiuchi, Tara Mitra, and Haley Bosco Doyle.

It is difficult to articulate the depth of love and companionship that Tish Lopez has so consistently offered me over the last ten years. She has accompanied me into so many places I didn't want to go. Our daily care for one another—even across great distances—and her heart-centeredness has sustained me through so much darkness.

To the members of my family: my mother, Anne Franks, who has been a constant source of love, support, and inspiration and who read earlier drafts of this book with such patience and insight; to my father, Peter Gillespie, for teaching me how to write, for working so tirelessly with me on finding the form and tone of this book, for making this process *fun*, and for coming along with me to the farm and the auction yard. Thank you to my sister, Lucy Gillespie, and to Ruth Saks and Chuck Sawyer for their support, care, and love.

To the nonhuman animals with whom my life has been so thoroughly and joyfully entwined — some still living and some now gone: George, Charlotte, Emily, and Jane; Abigail and Eden; Saoirse, Lucy, and Amelia; Maizy and Mally — they have taught me how to love and care in a whole new way.

Eric Haberman — my partner, my love, and my best friend — has lived this work every day with me. His unflinching love, patience, material and emotional support, and humor have made it possible for me to learn and grow and to do what I am most passionate about in life. There is no greater gift.

To the animals and their ghosts who inhabit these pages, I am always in their debt as some of my greatest teachers. Their lives, labor, suffering, grief, and deaths are not forgotten. May these pages and the things that grow from them ease the plight of those who come after them.

NOTES

Chapter 1

1. See Jeffrey Moussaieff Masson's *The Pig Who Sang to the Moon: The Emotional World of Farm Animals* (New York: Ballantine Books, 2004), and *The Face on Your Plate: The Truth about Food* (New York: W.W. Norton & Company, 2009); and Barbara King's *The Personalities on Your Plate* (Chicago: University of Chicago Press, 2017).

2. Marc Bekoff, "Animal Emotions: Exploring Passionate Natures," *BioScience* 50, no. 10 (2000): 867.

3. See Marc Bekoff's *The Emotional Lives of Animals* (Novato, CA: New World Library, 2007); Jane Goodall's *The Chimpanzees of Gombe* (Cambridge, MA: Belknap Press, 1986); Barbara King's *How Animals Grieve* (Chicago: University of Chicago Press, 2013), and *The Personalities on Your Plate*.

4. Lori Gruen refers to this as a kind of "caring perception," and explores how entangled empathetic engagement helps us to understand individuals of other species in her book *Entangled Empathy: An Alternative Ethic for Our Relationship with Animals* (New York: Lantern Books, 2014).

5. The nutria is a semiaquatic rodent native to South America that looks like a large river rat or like a beaver with a ratlike tail. Called coypu in South America, the species was introduced to North America and other parts of the world by the fur trade.

6. Michael Parenti, "Lies, War and Empire," lecture given at Antioch University, Seattle, WA, May 12, 2007, video, 1:28:28, published on September 9, 2012, http://www.youtube.com/watch?v=Rt_iAXYBUSk.

7. Joan Dunayer's *Animal Equality: Language and Liberation* (Derwood, MD: Ryce Publishing, 2001), and *Speciesism* (Derwood, MD: Ryce Publishing, 2004) are central texts dedicated to understanding the way language operates to maintain hierarchies of power and dominance.

8. For further reading on the commodity, see Karl Marx's *Capital* (1867; repr., Moscow: Progress Press, 1965); David Harvey's *A Companion to Marx's Capital* (London: Verso, 2010); and Arjun Appadurai *The Social Life of Things* (Cambridge: Cambridge University Press, 1986).

9. Rosemary-Claire Collard and Jessica Dempsey, "Life for Sale? The Politics of Lively Commodities," *Environment and Planning A* 45 (2013): 2864.

10. See YouTube channel RealCaliforniaMilk (https://www.youtube.com/user/Real

CaliforniaMilk) or visit https://www.youtube.com/watch?v=VbNf8p63aVE for an example of a Real California Milk commercial, titled "Alarm Clock."

11. See, for example, Melanie DuPuis's *Nature's Perfect Food* (New York: New York University Press, 2002); or Deborah Valenze's *Milk: A Local and Global History* (New Haven, CT: Yale University Press, 2011).

12. Wendy LiKamWa McIntosh, Erica Spies, Deborah M. Stone, Colby N. Lokey, Aimée-Rika T. Trudeau, Brad Bartholow, "Suicide Rates by Occupational Group—17 States, 2012," *Morbidity and Mortality Weekly Report*, 65, no. 25 (2016): 641–45, https://www.cdc.gov/mmwr/volumes/65/wr/pdfs/mm6525.pdf.

13. Robin R. Ganzert, *Humane Heartland™ Farm Animal Welfare Survey* (Washington, DC: American Humane Association, 2013), https://www.americanhumane.org/app/uploads/2016/08/2014-humane-heartland-farm-survey.pdf.

14. Jayson L. Lusk, F. Bailey Norwood, and Robert W. Prickett, "Consumer Preferences for Farm Animal Welfare: Results of a Nationwide Telephone Survey" (working paper, Department of Agricultural Economics, Oklahoma State University, 2007), http://cratefreefuture.com/pdf/American%20Farm%20Bureau-Funded%20Poll.pdf.

15. Jonathan Safran Foer, *Eating Animals* (New York: Little, Brown and Company, 2009), 55–56.

16. Milk Truth, "It's Time to Get Real about Milk," Milk Processors Education Program, 2015, accessed February 23, 2017, http://milktruth.com/.

17. Marion Nestle, *Food Politics* (Berkeley: University of California Press, 2007), 41–42.

18. USDA, "MyPlate," accessed February 23, 2017, https://www.choosemyplate.gov/.

19. Milk Truth, "It's Time to Get Real about Milk," Milk Processors Education Program, 2015, accessed February 23, 2017, http://milktruth.com/.

20. USDA, National Agricultural Statistical Service, "Data Set: Milk Cows and Production by State and Region," US Department of Agriculture: Economic Research Service, last updated May 11, 2017, https://www.ers.usda.gov/data-products/dairy-data/.

21. USDA, National Agricultural Statistical Service, "Data Set: Livestock and Poultry Slaughter, Historical," US Department of Agriculture: Economic Research Service, last updated January 30, 2018, https://www.ers.usda.gov/data-products/livestock-meat-domestic-data/. Note: these numbers exclude animals slaughtered on farms and reflect only animals slaughtered in federally inspected facilities.

22. David J. Wolfson, "Beyond the Law: Agribusiness and the Systemic Abuse of Animals Raised for Food or Food Production," *Lewis and Clark Animal Law Review* 2 (1995–96): 4, accessed February 17, 2014, http://www.animallaw.info/articles/arus2anima1123.htm.

23. James M. MacDonald, Erik J. O'Donoghue, William D. McBride, Richard F. Nehring, Carmen L. Sandretto, and Roberto Mosheim, *Profits, Costs and the Changing Structure of Dairy Farming*, Economic Research Report, no. 47, September 2007, https://www.ers.usda.gov/webdocs/publications/45868/11138_err47_1_.pdf?v=41746, 2.

24. USDA, *Overview of the United States Dairy Industry*, released September 22, 2010, by the National Agricultural Statistics Service (NASS), Agricultural Statistics Board, USDA, http://usda.mannlib.cornell.edu/usda/current/USDairyIndus/USDairyIndus-09-22-2010.pdf, 1.

25. USDA, *Overview of the United States Dairy Industry*, 1.

26. MacDonald et al., *Profits, Costs and the Changing Structure of Dairy Farming*, 2.

27. USDA, *Overview of the United States Dairy Industry*, 1; and MacDonald et al., *Profits, Costs and the Changing Structure of Dairy Farming*, 2.

28. For more on this process of consolidating and intensifying food production, see William Cronon's *Nature's Metropolis* (New York: W. W. Norton, 1991); Wendell Berry's *The Unsettling of America* (Berkeley, CA: Sierra Club Books, 1997); Eric Schlosser's *Fast Food Nation* (Boston: Houghton Mifflin, 2001); and Tony Weis's *The Global Food Economy* (New York: Zed Books, 2007).

29. USDA, *Overview of the United States Dairy Industry*, 9.

30. USDA, *Overview of the United States Dairy Industry*, 9.

31. USDA, *Overview of the United States Dairy Industry*, 9.

32. To provide a bit more detail on my field research process: I visited the Washington dairy farm in June of 2012, spent roughly three hours there, interviewing long-time farmer Homer Weston and touring the farm. I spent a full day and a half (one overnight) at Farm Sanctuary in Chico, CA, in June 2012, toured the sanctuary, observed the animals, and interviewed four employees there (two animal caregivers, and two education-focused employees); these interviews ranged from one to two hours each. Immediately following this visit, I spent four hours at Animal Place in Grass Valley, CA, interviewing the education director and touring the sanctuary. I attended eight different auctions at three different auction yards (two auction yards in Washington and one in California) between June 2012 and December 2012, spending between two and three hours at each auction (and exploring the auction grounds). I attended the World Dairy Expo for one full day in October 2012 in Madison, WI. I conducted four interviews of between one and two hours each with adults who had been involved in 4-H as children (these interviews were conducted between August 2012 and January 2013, and I found interviewees through a snowball sampling technique, where people I knew referred me to others they knew who had been involved in 4-H). I attended the Washington State Fair in Puyallup, WA, in September 2012 and spent six hours walking the fair grounds and observing, especially, the 4-H dairy event and animal areas. My textual and discourse analysis of published texts and online materials spanned from September 2011 to August 2013.

33. Donna Haraway, "Situated Knowledges: The Science Question in Feminism and the Privilege of Partial Perspective," *Feminist Studies* 14, no. 3 (1988): 577–99.

34. See Lori Gruen's *Entangled Empathy* (New York: Lantern Books, 2014); Kathryn Gillespie, "Intimacy, Animal Emotion, and Empathy: Multispecies Intimacy as Slow Research Practice," in *Writing Intimacy into Feminist Geography*, ed. Pamela Moss and Courtney Donovan (London: Routledge, 2017), and "Witnessing Animal Others: Bearing Witness, Grief, and the Political Function of Emotion," *Hypatia* 31, no. 3 (2016): 572–88.

Chapter 2

1. Food and Drug Administration (FDA), *2011 Summary Report on Antimicrobials Sold or Distributed for Use in Food-Producing Animals*, FDA, Department of Health and Human Services, September 2014, http://www.fda.gov/downloads/ForIndustry/UserFees/AnimalDrugUserFeeActADUFA/UCM338170.pdf.

2. Centers for Disease Control and Prevention (CDC), *Antibiotic Resistance Threats in the United States, 2013*, US Department of Health and Human Services, CDC, 2013, accessed March 18, 2015, http://www.cdc.gov/drugresistance/pdf/ar-threats-2013-508.pdf.

3. Animals termed *downers* by the industry are those who have become nonambulatory and cannot move on their own. The cow with ear tag #1389, for example, may have been characterized as a downer cow, in spite of the fact that signage at the auction yard declared that they did not allow nonambulatory animals into the auction.

4. Will Potter, "First 'Ag-Gag' Prosecution," *Green Is the New Red* (blog), April 29, 2013, http://www.greenisthenewred.com/blog/first-ag-gag-arrest-utah-amy-meyer/6948/.

5. Humane Society of the United States, "Owners of Infamous Hallmark Meat Company Pay $300,000 in HSUS Slaughterhouse Cruelty Case," press release, November 16, 2012, http://www.humanesociety.org/news/press_releases/2012/11/hallmark-meat-company-settlement-111612.html.

6. See Will Potter, "Filming This Slaughterhouse from the Street was the First 'Ag-Gag' Prosecution," film footage recorded by Amy Meyer, 2013, video, 9:37, published on June 24, 2013, http://www.youtube.com/watch?feature=player_embedded&v=9HIsA8EIWkQ.

7. Providing Legal Protection to Animal Owners and Their Animals and to Ensure That Only Law Enforcement Agencies Investigate Charges of Animal Cruelty, AR SB 13, (passed April 12, 2013), https://legiscan.com/AR/text/SB13/2013.

8. Iowa House File 589: An Act Relating to an Offense Involving Agricultural Operations, and Providing Penalties, and Including Effective Date Provisions," HF 589 (passed March 2, 2012), http://coolice.legis.iowa.gov/Cool-ICE/default.asp?Category=billinfo&Service=Billbook&menu=false&ga=84&hbill=HF589.

9. SB 631: Modified Provisions Relating to Animals and Agriculture, SB 631 (passed July 9, 2012), http://www.senate.mo.gov/12info/BTS_Web/Bill.aspx?SessionType=R&BillID=92863.

10. Agricultural Operation Interference, HB 187, 2012, http://le.utah.gov/~2012/bills/hbillint/hb0187.htm.

11. Potter, "First 'Ag-Gag' Prosecution."

12. Will Potter, *Green Is the New Red*, (San Francisco: City Lights, 2011).

13. Will Potter, "The Green Scare," *Vermont Law Review* 33, no. 4 (2009): 672–73.

14. Federal Bureau of Investigations, "General Functions," 28 C.F.R. sec. 0.85, July 1, 2010, http://www.gpo.gov/fdsys/pkg/CFR-2010-title28-vol1/pdf/CFR-2010-title28-vol1-seco-85.pdf.

15. USA PATRIOT Act, Pub. L. No. 107-56, sec. 802 (2001), http://www.gpo.gov/fdsys/pkg/PLAW-107pub156/pdf/PLAW-107pub156.pdf.

16. Jerome P. Bjelopera, *The Domestic Terrorist Threat: Background and Issues for Congress*, Congressional Research Service, CRS Report to Congress, January 17, 2013, http://fas.org/sgp/crs/terror/R42536.pdf.

17. There are interesting questions to be asked about the nature of what constitutes "terrorism" in the context of the AETA and ag-gag laws. Many of these questions are explored in the thought-provoking documentary, *If a Tree Falls: A Story of the Earth Liberation Front* (directed by Marshall Curry and Sam Cullman [Brooklyn, NY: Oscilloscope Pictures, 2011], DVD), which follows the case of a series of arsons committed by members

of the Earth Liberation Front in the Pacific Northwest. Exploring various sides of the issue, the documentary interviews former activists from the organization, community members, law enforcement, and the people behind the industries that were the target of the organization's direct actions.

18. For information on Fair Oaks Farms and its tours, see their website at http://fofarms .com/; Jan Dutkiewicz, "Transparency and the Factory Farm: Agritourism and Counter-Activism at Fair Oaks Farms," *Gastronomica* 18, no. 2 (2018): 19–32; Timothy Pachirat, "Are We All Phalaris Now? Pleasure, Pain, and (In)visible Suffering in Our Modern Times" (unpublished manuscript).

19. For example, Washington State law states that:

> (1) A person is guilty of animal cruelty in the first degree when, except as autho-rized in law, he or she intentionally (a) inflicts substantial pain on, (b) causes physical injury to, or (c) kills an animal by a means causing undue suffering, or forces a minor to inflict unnecessary pain, injury, or death on an animal.
>
> (2) A person is guilty of animal cruelty in the first degree when, except as authorized by law, he or she, with criminal negligence, starves, dehydrates, or suffocates an animal and as a result causes: (a) Substantial and unjustifiable physical pain that extends for a period sufficient to cause considerable suffer-ing; or (b) death.

Washington State Legislature, RCW 16.52.205, 2010, http://apps.leg.wa.gov/RCW /default.aspx?cite=16.52.205.

20. David J. Wolfson and Mariann Sullivan, "Foxes in the Hen House: Animals, Agribusi-ness, and the Law: A Modern American Fable," in *Animal Rights: Current Debates and New Directions*, ed. Cass R. Sunstein and Martha Nussbaum (Oxford: Oxford University Press, 2005), 209.

21. Wolfson and Sullivan, "Foxes in the Hen House," 210.

22. Erik Marcus, *Meat Market: Animals, Ethics and Money* (Ithaca, NY: Brio Press, 2005), 57.

23. Wolfson and Sullivan, "Foxes in the Hen House," 206.

24. Washington State Legislature, "Exclusions from Chapter," RCW 16.52.185, 2010, http:// apps.leg.wa.gov/RCW/default.aspx?cite=16.52.185.

25. Will Kymlicka and Sue Donaldson, "Animal Rights, Multiculturalism, and the Left," *Journal of Social Philosophy* 45, no. 1 (2014): 116–35, esp. 126, 132.

26. See Claire Jean Kim's *Dangerous Crossings: Race, Species, and Nature in a Multicultural Age* (Cambridge: Cambridge University Press, 2015).

27. Sophie Williams, "No, the Yulin Dog Meat Festival Has Not Been Cancelled —Here's Why It Will Never Go Away," *Independent*, June 21, 2017, http://www.independent .co.uk/voices/yulin-dog-meat-festival-china-animal-rights-chinese-culture-western -interference-a7800416.html; Humane Society International, "Ending the Yulin Dog Meat 'Festival,'" accessed February 18, 2018, http://www.hsi.org/issues/dog_meat /facts/stopping-yulin-festival.html.

28. Kymlicka and Donaldson, "Animal Rights, Multiculturalism, and the Left," 122; Manee-sha Deckha, "Animal Justice, Cultural Justice: A Posthumanist Response to Cultural Rights in Animals," *Journal of Animal Law and Ethics* 2 (2007): 189–229.

29. Maneesha Deckha, "Initiating a Non-anthropocentric Jurisprudence: The Rule of Law and Animal Vulnerability under a Property Paradigm," *Alberta Law Review* 50, no. 4 (2013): 783–814.

Chapter 3

1.   Ansel Farm and Homer Weston are both pseudonyms.

2.   A burdizzo is a pliers-like device used to castrate farmed animals. It clamps and severs the blood vessels leading to the testicles, which, over time, will shrivel and disappear. This tool can also be used as an improvised tail-docking tool.

3.   Animal Welfare Division, "Welfare Implications of Tail Docking of Cattle: Literature Review," American Veterinary Medical Association, August 29, 2014, https://www .avma.org/KB/Resources/LiteratureReviews/Pages/Welfare-Implications-of-Tail -Docking-of-Cattle.aspx.

4.   S. D. Eicher, H. W. Cheng, A. D. Sorrells, and M. M. Schutz, "Short Communication: Behavioral and Physiological Indicators of Sensitivity or Chronic Pain Following Tail Docking," *Journal of Dairy Science* 89, no. 8 (2006): 3047–51.

5.   Canadian Veterinary Medical Association, "Tail Docking of Dairy Cattle—Position Statement," October 12, 2016, https://www.canadianveterinarians.net/documents/tail -docking-of-dairy-cattle.

6.   "Tail Docking of Cattle," American Veterinary Medical Association Policies, August 29, 2014, https://www.avma.org/KB/Policies/Pages/Tail-Docking-of-Cattle.aspx.

7.   For more on the impacts of separation of cow and calf, see Frances C. Flower and Daniel M. Weary, "Effects of Early Separation on the Dairy Cow and Calf," *Applied Animal Behaviour Science* 70, no. 4 (2001): 275–84.

8.   *Dairy 2014: Dairy Cattle Management Practices in the United States, 2014*, USDA, Animal and Plant Health Inspection Service, Report 1, February 2016, https://www.aphis.usda .gov/animal_health/nahms/dairy/downloads/dairy14/Dairy14_dr_PartI.pdf.

9.   Purdue University School of Agriculture, "Dairy Production: Diseases," in *Ag 101* (Washington, DC: Environmental Protection Agency, 2002), https://www.epa.gov /sites/production/files/2015-07/documents/ag_101_agriculture_us_epa_0.pdf.

10. Mel DeJarnette and Ray Nebel, *A.I. Technique in Cattle* (Plain City, OH: Select Sires, 2012), accessed on December 4, 2014, http://www.selectsires.com/resources /fertilitydocs/ai_technique_cattle.pdf.

11. Jim Paulson et al., *Learning about Dairy*, Regents of the University of Minnesota, University of Minnesota Extension, rev. November 2015, http://www.extension.umn .edu/youth/mn4-H/events/project-bowl/docs/PB-Learning-About-Dairy-Booklet .pdf.

12. Paulson et al., *Learning about Dairy*.

13. Paulson et al., *Learning about Dairy*.

14. Animal Welfare Division, "Welfare Implications of Dehorning and Disbudding Cattle: Literature Review," American Veterinary Medical Association, July 15, 2014, https:// www.avma.org/KB/Resources/LiteratureReviews/Pages/Welfare-Implications-of -Dehorning-and-Disbudding-Cattle.aspx.

15. Fred M. Hopkins, James B. Neel, and F. David Kirkpatrick, *Dehorning Calves*, Uni-

versity of Tennessee, Agricultural Extension Service, accessed August 5, 2015, https://utextension.tennessee.edu/publications/documents/pb1684.pdf.

16. Animal Welfare Division, "Welfare Implications of Dehorning and Disbudding Cattle."

17. See Michael Pollan's *The Omnivore's Dilemma* (New York: Penguin Books, 2007) for a discussion of the transition to corn as a feed crop for farmed animals.

18. Lewis Holloway and Christopher Bear, "Bovine and Human Becomings in Histories of Dairy Technologies: Robotic Milking Systems and Remaking Animal and Human Subjectivity," *BJHS Themes* 2 (2017): 215–34; Lewis Holloway, Christopher Bear, and Katy Wilkinson, "Re-capturing Bovine Life: Robot-Cow Relationships, Freedom and Control in Dairy Farming," *Journal of Rural Studies* 33, no. 1 (2014): 131–40.

19. Kelsey Gee, "America's Dairy Farmers Dump 43 Million Gallons of Excess Milk," *Wall Street Journal*, October 12, 2016, https://www.wsj.com/articles/americas-dairy-farmers-dump-43-million-gallons-of-excess-milk-1476284353

20. Heather Haddon, "Got Milk? Too Much Of It, Say U.S. Dairy Farmers," *MarketWatch*, May 21, 2017, https://www.marketwatch.com/story/got-milk-too-much-of-it-say-us-dairy-farmers-2017-05-21

21. Gee, "America's Dairy Farmers."

22. Gee, "America's Dairy Farmers"; Haddon, "Got Milk?"

23. Drew Atkins, "Fecal Matter Pollution in Public Water," *Crosscut*, November 4, 2015, http://crosscut.com/2015/11/fecal-matter-pollution-in-drinking-water-the-case-of-snydar-farm/.

24. Elizabeth Johnson, "Governing Jellyfish: Eco-Security and Planetary 'Life' in the Anthropocene," in *Animals, Biopolitics, Law: Lively Legalities*, ed. Irus Braverman (London: Routledge, 2016).

25. Feminist critical animal studies scholars have long noted the effects of gendered commodification and violence against the animal body with particular emphasis on the ways in which female animals are disproportionately exploited for their (re)productive capabilities in places of commodity production. For examples, see Carol J. Adams's *The Sexual Politics of Meat* (New York: Continuum, 1990); Lori Gruen's "Dismantling Oppression," in *Ecofeminism*, ed. Greta Gaard (Philadelphia: Temple University Press, 1993); and Carol J. Adams and Josephine Donovan, eds., *Animals and Women* (Durham, NC: Duke University Press, 1995).

26. On cage size, see United Egg Producers, *United Egg Producers Animal Husbandry Guidelines for US Egg Laying Flocks: 2016 Edition* (Alpharetta, GA: United Egg Producers, 2010), https://uepcertified.com/wp-content/uploads/2015/08/UEP-Animal-Welfare-Guidelines-20141.pdf. And regarding debeaking, see Heng-Wei Cheng, *Genetic Selection and Welfare in Laying Hens*, Laying Hen Welfare Fact Sheet, USDA Livestock Behavior Research Unit, Summer 2011, http://www.ars.usda.gov/SP2UserFiles/Place/36022000/Genetic%20Selection%20Fact%20Sheet.pdf.

27. Steven Wise, *An American Trilogy: Death, Slavery and Dominion on the Cape Fear River* (Philadelphia: Da Capo Press, 2009).

28. María Elena García, "Super Guinea Pigs," *Anthropology Now* 2, no. 2 (2010): 22–32, and "The Taste of Conquest: Colonialism, Cosmopolitics, and the Dark Side of Peru's Gastronomic Boom," *Journal of Latin American and Caribbean Anthropology* 18, no. 3: 505–24.

Chapter 4

1. E. M. C. Terlouw, C. Arnould, B. Auperin, C. Berri, E. Le Bihan-Duval, V. Deiss, F. Lefèvre, B. J. Lensink, and L. Mounier, "Pre-slaughter Conditions, Animal Stress and Welfare: Current Status and Possible Future Research," *Animal* 2, no. 10 (2008): 1501–17.

2. Kathryn Gillespie. "Nonhuman Animal Resistance and the Improprieties of Live Property," in *Animals, Biopolitics, Law: Lively Legalities*, ed. Irus Braverman (London: Routledge, 2016).

3. Animal Welfare Division, "Welfare Implications of Hot-Iron Branding and Its Alternatives: Literature Review," American Veterinary Medical Association, April 4, 2011, https://www.avma.org/KB/Resources/LiteratureReviews/Pages/Welfare-Implications-of-Hot-Iron-Branding-and-Its-Alternatives.aspx.

4. Ralph Cassady, *Auctions and Auctioneering* (Berkeley: University of California Press, 1967).

5. Cassady, *Auctions and Auctioneering*, 90.

6. Rosemary-Claire Collard, in her book *Animal Traffic* (Durham, NC: Duke University Press, forthcoming), describes this act of severing in great detail in the context of capturing free-living animals from their wild habitats for commodification in the global exotic pet trade. She describes, for instance, the capture of infant spider monkeys, accomplished by shooting the adult out of the trees and removing the infant as they cling to the adult's dead body. Or the removal of baby macaws from their nests high in the forest canopy. These acts of severing are deeply troubling, especially as they span wild and domesticated contexts.

Chapter 5

1. Patricia J. Lopez and Kathryn Gillespie, "A Love Story: For 'Buddy System' Research in the Academy," *Gender, Place, and Culture* 23, no. 12 (2016): 1689–1700.

2. Kathryn Gillespie, "Witnessing Animal Others: Bearing Witness, Grief, and the Political Function of Emotion." *Hypatia* 31, no. 3 (2016): 572–88.

3. See Leonardo Nonni Costa, "Short-term Stress: The Case of Transport and Slaughter," *Italian Journal of Animal Science* 8, no. 1 (2009): 241–52; E. M. C. Terlouw, C. Arnould, B. Auperin, C. Berri, E. Le Bihan-Duval, V. Deiss, F. Lefèvre, B. J. Lensink, and L. Mounier, "Pre-slaughter Conditions, Animal Stress and Welfare: Current Status and Possible Future Research." *Animal* 2, no. 10 (2008): 1501–17.

4. See P. M. Sihom, "Welfare of Cattle Transported from Australia to Egypt," *Australian Veterinary Journal* 81 (2003): 364; N. G. Gregory, "Animal Welfare at Markets and during Transport and Slaughter," *Meat Science* 80, no. 1 (2008): 2–11.

5. Terlouw et al., "Pre-slaughter Conditions, Animal Stress and Welfare."

6. Lucinda Holt, "Cattle Trailer Collision with Tractor-Trailer Injures Driver; Dozens of Animals Dead or Hurt," *Lubbock Avalanche Journal*, November 3, 2016, http://lubbockonline.com/filed-online/2016–11–03/cattle-trailer-collision-tractor-trailer-injures-driver-dozens-animals-dead.

7.   Lauren O'Neil, "Thousands of Piglets Run Free in Ohio after Truck Crashes on High-way," *CBC News*, June 9, 2015, http://www.cbc.ca/news/trending/thousands-of-piglets-run-free-in-ohio-after-truck-crashes-on-highway-1.3106641.

8.   Steve Striffler, *Chicken: The Dangerous Transformation of America's Favorite Food* (New Haven, CT: Yale University Press, 2005).

9.   Timothy Pachirat, *Every Twelve Seconds: Industrialized Slaughter and the Politics of Sight* (New Haven, CT: Yale University Press, 2011).

10. Temple Grandin, *Animals Make Us Human: Creating the Best Life for Animals* (New York: Mariner Books, 2010).

11. Pachirat, *Every Twelve Seconds*, chap. 9

12. Pachirat, *Every Twelve Seconds*, 253.

13. US Bureau of Labor Statistics. "Household Data Annual Averages," accessed January 7, 2016, http://www.bls.gov/cps/cpsaat11.pdf.

14. Mobile Slaughter Unit, accessed January 6, 2017, http://mobileslaughter.com/.

15. Jared Diamond, *Guns, Germs, and Steel* (New York: W.W. Norton & Company, 1999).

16. Kevin and Urban Rendering are pseudonyms.

17. Dave is a pseudonym.

18. Information about Darling International, can be found at their website: https://www.darlingii.com/.

19. Kurt is a pseudonym.

20. For an excellent cultural politics analysis of rendering, see Nicole Shukin's *Animal Capital: Rendering Life in Biopolitical Times* (Minneapolis: University of Minnesota Press, 2009).

21. National Renderers Association, "Rendering Is Recycling," accessed June 22, 2015, https://d1ok7k7mywg422.cloudfront.net/assets/53e623d14f720a3623000255/NRA_infographic_ONE_PAGE_web_01.jpg.

22. David L. Meeker and C. R. Hamilton, "An Overview of the Rendering Industry," National Renderers Association, accessed January 31, 2018, http://assets.nationalrenderers.org/essential_rendering_overview.pdf.

23. National Renderers Association, "Rendering Is Recycling."

24. Meeker and Hamilton, "An Overview of the Rendering Industry," 3.

25. Information from this paragraph is based on Meeker and Hamilton, "An Overview of the Rendering Industry."

26. Meeker and Hamilton, "An Overview of the Rendering Industry."

Chapter 6

1.   Ryan is a pseudonym.

2.   Sue Donaldson and Will Kymlicka, "Farmed Animal Sanctuaries: The Heart of the Movement? A Socio-Political Perspective," *Politics and Animals* 1 (2015): 54.

3.   Elan Abrell, "Saving Animals: Everyday Practices of Care and Rescue in the US Animal Sanctuary Movement," (PhD diss., City University of New York, 2016).

4.   pattrice jones, "Afterword: Flower Power," in *Confronting Animal Exploitation: Grassroots Essays on Liberation and Veganism*, ed. K. Socha and S. Blum (Jefferson, NC:

McFarland, 2013); Norm Phelps, *Changing the Game: Why the Battle for Animal Liberation Is So Hard and How We Can Win It* (New York: Lantern, 2015); Donaldson and Kymlicka, "Farmed Animal Sanctuaries."

5. Anne is a pseudonym.
6. USDA, "Milk Production," USDA National Agricultural Statistics Service, released February 21, 2017, http://usda.mannlib.cornell.edu/usda/nass/MilkProd//2010s/2017/MilkProd-02-21-2017.pdf, 9.
7. USDA, "Milk Production on December 1, 1931," USDA Bureau of Agricultural Economics, December 18, 1931, http://usda.mannlib.cornell.edu/usda/nass/MilkProd//1930s/1931/MilkProd-12-18-1931.pdf, 2; USDA, "Milk Production," released February 21, 2017, 1.
8. Donaldson and Kymlicka, "Farmed Animal Sanctuaries."
9. Donaldson and Kymlicka, "Farmed Animal Sanctuaries," 57.
10. Donaldson and Kymlicka, "Farmed Animal Sanctuaries," 57.
11. pattrice jones, *The Oxen at the Intersection: A Collision* (New York: Lantern Books, 2014), 73.
12. Abrell, "Saving Animals," vii.
13. Julie is a pseudonym.
14. Sunaura Taylor, *Beasts of Burden: Animal and Disability Liberation* (New York: New Press, 2017).
15. Eli Clare, *Brilliant Imperfection: Grappling with Cure* (Durham, NC: Duke University Press, 2017), 26.
16. For a more extended discussion of reproductive control at sanctuaries, see Donaldson and Kymlicka, "Farmed Animal Sanctuaries," 57.
17. Miriam Jones, "Captivity in the Context of a Sanctuary for Formerly Farmed Animals," in *The Ethics of Captivity*, ed. Lori Gruen (Oxford: Oxford University Press, 2014), 91–92.
18. For more information on veganic agriculture, you can check out the Veganic Agriculture Network online (https://www.goveganic.net/), which promotes plant-based farming and gardening throughout North America.
19. As of the writing of this book, Beach had informed me that the vegan micro farm project had ended when the professor who ran the project moved out of state.
20. Many thanks to Lori Gruen for her thinking on the many ways that neglect, abuse, and cruelty can occur—namely, that neglect and cruelty manifest in highly varied relationships of care.

Chapter 7

1. George Orwell, *1984*, accessed February 18, 2018, https://www.planetebook.com/ebooks/1984.pdf, 44–45.
2. No one is born ignorant; ignorance is a derivative state of mind produced by ignoring; ignorance is something you develop and grow into. You can't ignore something unless you are aware of it and have some motive for denying that awareness. Ignorance is not the same as being oblivious; though the word *ig-norant* from its roots says literally *not-knowing*, it's not a simple not knowing. What is called *ig-noring* entails a knowing (the

"-noring" part comes from the Latin verb *noscere*—to know) which is negated by the prefix *ig-* which says *not*.

3. Allie Novak is a pseudonym.

4. "4-H History: The Birth of 4-H Programs," 4-H, 2013, accessed June 30, 2015, http://www.4-h.org/about/4-h-history/.

5. "Dairy Cattle: Cowabunga," 4-H, accessed July 1, 2015, http://www.4-h.org/resource-library/curriculum/4-h-dairy-cattle/cowabunga/.

6. "Dairy Cattle: Mooving Ahead," 4-H, accessed July 1, 2015, http://www.4-h.org/resource-library(curriculum/4-h-dairy-cattle/mooving-ahead/.

7. "Dairy Cattle: Rising to the Top," 4-H, accessed July 1, 2015, http://www.4-h.org/resource-library/curriculum/4-h-dairy-cattle/rising-to-the-top/.

8. Deborah Y. Richardson, *All about Dairy Cows* (Beltsville, MD: US Department of Agriculture, 2003), http://www.4-h.org/resource-library/curriculum/4-h-dairy-cattle/cowabunga/, 6.

9. Richardson, *All about Dairy Cows,* 10.

10. Richardson, *All about Dairy Cows,* 11.

11. Richardson, *All about Dairy Cows,* 19.

12. "Ethics in the Show Ring: Making the Responsible Choice" *Industry Image,* July 2000, reprinted from *Jersey Journal,* http://www.4-h.org/resource-library/curriculum/4-h-dairy-cattle/mooving-ahead/.

13. James Connors and Janice Dever, "Unethical Practices Observed at Youth Livestock Exhibitions by Ohio Secondary Agricultural Educators," *Journal of Agricultural Education* 46, no. 1 (2005): 20–31. Joanne L. Goodwin, "The Ethics of Livestock Shows—Past, Present, and Future," *Journal of the American Veterinary Medical Association* 219 (2001): 1391–93.

14. Connors and Dever, "Unethical Practices Observed."

15. Connors and Dever, "Unethical Practices Observed."

16. "Ethics in the Show Ring," 1.

17. For a discussion of the concept of animals as "subjects of a life," see Tom Regan's *The Case for Animal Rights* (Berkeley: University of California Press, 2004).

18. Heidi Sloane is a pseudonym.

19. Dana Gunders, "Wasted: How America Is Losing Up to 40 Percent of Its Food from Farm to Fork to Landfill," National Resources Defense Council, NRDC Issue Paper, August 2012, http://www.nrdc.org/food/files/wasted-food-IP.pdf.

20. For an example of this phenomenon in the Pacific Northwest, see Victoria Lawson, Lucy Jarosz, and Anne Bonds, "Articulations of Place, Poverty, and Race: Dumping Grounds and Unseen Grounds in the Rural American Northwest," *Annals of the Association of American Geographers* 100, no. 3 (2010): 655–77.

21. For more on human-pet relations, see Yi-Fu Tuan's *Dominance and Affection* (New Haven, CT: Yale University Press, 2004); and Jessica Pierce's *Run, Spot, Run: The Ethics of Keeping Pets* (Chicago: University of Chicago Press, 2016).

22. See Lori Gruen's *Entangled Empathy* (New York: Lantern Books, 2015).

Chapter 8

1. Information on the World Dairy Expo can be found at their website: http://worlddairyexpo.com/.

2. Sara Kober, "In Vitro Fertilization and Embryo Transfer: A Comparison," Trans Ova Genetics blog, July 7, 2017, http://www.transova.com/tog-blog/in-vitro-fertilization-embryo-transfer-a-comparison.

3. Carlos A. Risco, Fabio Lima, Jose E. P. Santos, "Management Considerations of Natural Service Breeding Programs in Dairy Herds," University of Florida Extension Program, accessed February 18, 2018, http://extension.vetmed.ufl.edu/files/2012/04/Management-Considerations-of-Natural-Service-Breeding-Programs-in-Dairy-Herds_Risco.pdf.

4. Risco et al., "Management Considerations," 1.

5. Risco et al., "Management Considerations," 1.

6. Risco et al., "Management Considerations," 1.

7. P. J. Chenoweth, J. D. Champaigne, and J. F. Smith, "Managing Herd Bulls on Large Dairies," *Proceedings of the Sixth Western Dairy Management Conference*, Reno, NV, March 12–14, 2003, http://www.asi.ksu.edu/doc4130.ashx, 107–18.

8. Temple Grandin, "Preventing Bull Accidents," grandin.com, June 2006, http://www.grandin.com/behaviour/principles/preventing.bull.accidents.html.

9. Animal Welfare Division, "Welfare Implications of Castration of Cattle: Literature Review," American Veterinary Medical Association, July 15, 2015, accessed August 5, 2015, https://www.avma.org/KB/Resources/LiteratureReviews/Pages/castration-cattle-bgnd.aspx.

10. Animal Welfare Division, "Welfare Implications of Castration of Cattle."

11. Animal Welfare Division, "Welfare Implications of Castration of Cattle."

12. Clyde Lane Jr., Richard Powell, Brian White, and Steve Glass, "Castration of Beef Calves," University of Tennessee Extension, Tennessee Agricultural Enhancement Program, document SP692, accessed August 5, 2015, https://utextension.tennessee.edu/publications/documents/sp692.pdf.

13. Animal Welfare Division, "Welfare Implications of Castration of Cattle."

14. Animal Welfare Division, "Welfare Implications of Castration of Cattle."

15. Jane M. Morrell, "Artificial Insemination: Current and Future Trends," in *Artificial Insemination in Farm Animals*, ed. Milad Manafi (n.p.: InTech, 2011), http://www.intechopen.com/books/artificial-insemination-in-farm-animals/artificial-insemination-current-and-future-trends.

16. John Cothren, "What Are the Advantages of Artificial Insemination (AI) in Your Livestock Breeding Program?" Wilkes Extension Center, NC State University Cooperative Extension, updated June 29, 2016, http://wilkes.ces.ncsu.edu/2012/12/what-are-the-advantages-of-using-artificial-insemination-ai-in-your-livestock-breeding-program/.

17. Melissa Rouge, "Semen Collection from Bulls," in *Pathophysiology of the Reproductive System*, Colorado State University, last updated September 2, 2002, http://www.vivo.colostate.edu/hbooks/pathphys/reprod/semeneval/bull.html.

18. Cathy Tesnohlidek Mosman, "Ejaculation and Semen Collection," AG 534-J8, in *Ag 534: Zoology—Science of Animal Reproduction: Course Outline* (Moscow: University

of Idaho, Agricultural and Extension Education, nd), 334, http://www.uidaho.edu
/-/media/UIdaho-Responsive/Files/cals/departments/AEE/educators/AG-534
-Zoology-science-of-animal-reproduction.ashx.

19. Rouge, "Semen Collection from Bulls."

20. Melissa Rouge and R. Bowen, "Semen Collection," in *Pathophysiology of the Reproductive System*, Colorado State University, last updated on August 11, 2002), http://www
.vivo.colostate.edu/hbooks/pathphys/reprod/semeneval/collection.html.

21. These categorizations, and the association of sex and gender with reproduction, echo
norms about the human body and can work to reinforce the exclusion and erasure of
transgender, gender nonconforming, and intersex lives and bodies.

22. Michel Foucault, *The History of Sexuality*, vol. 1, trans. R. Hurley. (New York: Pantheon,
1978).

23. Donna Haraway, *Primate Visions* (London: Routledge, 1989), 289.

24. BouMatic, SmartDairy Activity Module, 2015, accessed February 18, 2018, http://
kruegersboumatic.com/wp-content/uploads/2015/08/SD_Actv_LIT00317EN-1203
_v6_EN.pdf.

25. Rovimix, DSM, Beta-Carotene, 2015, ad in "Market Watch," *Progressive Dairyman*, no. 9,
June 11, 2012, 2, https://www.progressivedairy.com/downloads/2012/06/0912pd_mw
_milk.pdf

26. Semex, "Put Time on Your Side," leaflet advertisement, accessed January 31, 2018,
http://www.semex.com/downloads/designer-us/ai24-2.pdf.

27. Cargill, "She's in It to Make Milk," advertisement, *Dairy Today*, September 2012 (also
available at https://krsharpe.files.wordpress.com/2014/12/kali-sharpe-cargill-ad.pdf).

28. Bovi-Shield Gold, "If she can't stay pregnant, what else will she do?" print advertisement, Pfizer Animal Health, 2012, retrieved from the World Dairy Expo.

29. The suggestion that the cow would have nothing to do if she didn't stay pregnant is a
reminder of the dominant social norms that a woman's job is to have children and reflects the pervasive discomfort about women who do not have children.

30. See Virginia Anderson's book *Creatures of Empire* (Oxford: Oxford University Press,
2006) for a more thorough history of the role of domesticated farmed animal species —
like cows and horses — in the settler-colonial project in the Americas.

31. Select Sires, "Superior Settlers," Select Sires Superior Settlers, catalog, 2012, retrieved
from the World Dairy Expo.

32. See Brian Luke, *Brutal: Manhood and the Exploitation of Animals* (Urbana: University
of Illinois Press, 2007).

33. Peter Robertson, *Pheasants* (London: Voyageur Press, Inc., 1997).

34. Cydectin ad, NuZu Feed, livestock catalog, 2, accessed February 26, 2018, http://www
.nuzufeed.com/Livestock2-75.pdf.

35. Select Sires, "Showcase Selections: Where Winning Begins," Select Sires advertising
catalog, 2012, retrieved from the World Dairy Expo.

36. Carol Adams's *Sexual Politics of Meat* (1990; repr., New York: Continuum, 2010) theorizes these connections in greater detail in addition to detailing many examples of the
connections between women and animals and sexualized imagery/discourses.

37. "Sammy Semen Clothing Collection," Universal Semen Sales, accessed August 3, 2015,
http://www.universalsemensales.com/sammy-semen-clothing-collection.

38.   Eli Clare, *Brilliant Imperfection: Grappling with Cure* (Durham, NC: Duke University Press, 2017), 134.

Chapter 9

1.   "Animal Manure Management," USDA, Natural Resources Conservation Service, RCA Issue Brief no. 7, December 1995, http://www.nrcs.usda.gov/wps/portal/nrcs/detail /national/technical/nra/?&cid=nrcs143_014211.

2.   "China Builds 100,000 Cow Dairy Farm," *Rotorua Daily Post*, August 13, 2015, http:// m.nzherald.co.nz/rotorua-daily-post/rural/news/article.cfm?c_id=1503433&objectid =11496705.

3.   Jim Paulson et al., *Learning about Dairy*, Regents of the University of Minnesota, University of Minnesota Extension, rev. November 2015, http://www.extension.umn .edu/youth/mn4-H/events/project-bowl/docs/PB-Learning-About-Dairy-Booklet .pdf, 79.

4.   "Product Water Footprint," Water Footprint Network, accessed April 7, 2015, http:// waterfootprint.org/en/water-footprint/product-water-footprint/.

5.   See Elizabeth Kolbert, *The Sixth Extinction* (New York: Picador, 2014).

6.   Henning Steinfeld, Pierre Gerber, Tom Wassenaar, Vincent Castel, Mauricio Rosales, and Cees de Haan, *Livestock's Long Shadow: Environmental Issues and Options* (Rome: Food and Agriculture Organization of the United Nations, 2006).

7.   *Cowspiracy: The Sustainability Secret*, directed by Kip Andersen and Keegan Kuhn (Santa Rosa, CA: A.U.M. Films & Media and First Spark Media, 2014), DVD.

8.   Purdue University School of Agriculture, "Dairy Production: Diseases," in *Ag 101* (Washington, DC: Environmental Protection Agency, 2002), https://www.epa.gov /sites/production/files/2015-07/documents/ag_101_agriculture_us_epa_0.pdf.

9.   Food Safety and Inspection Service, "Veal from Farm to Table," Food Safety and Inspection Service, USDA, last updated August 6, 2013, accessed July 5, 2015, http:// www.fsis.usda.gov/wps/portal/fsis/topics/food-safety-education/get-answers/food -safety-fact-sheets/meat-preparation/veal-from-farm-to-table/CT_Index.

10.   Amy Pennington, "Consider the Calf," *Edible Seattle*, July–August 2012, http:// edibleseattle.com/consider-the-calf/.

11.   See Tom Regan's *The Case for Animal Rights*.

12.   Rebekah Denn, "Sea Breeze Farm raises Veal the 'Real' Way—in Pastures," *Seattle Times*, August 30, 2009, https://www.seattletimes.com/pacific-nw-magazine/sea -breeze-farm-raises-veal-the-real-way-8212-in-pastures/.

13.   Bovine Alliance on Management and Nutrition, *A Guide to Colostrum and Colostrum Management for Dairy Calves*, USDA, APHIS, 2001, accessed August 5, 2015, http:// www.aphis.usda.gov/animal_health/nahms/dairy/downloads/bamn/BAMN01 _Colostrum.pdf.

14.   BAMN, *A Guide to Colostrum and Colostrum Management.*

15.   BAMN, *A Guide to Calf Milk Replacers*, USDA, APHIS, 2008, accessed August 5, 2015, http://www.aphis.usda.gov/animal_health/nahms/dairy/downloads/bamn /BAMN08_GuideMilkRepl.pdf.

16.   BAMN, *A Guide to Dairy Calf Feeding and Management*, USDA, APHIS, 2003, accessed

February 26, 2018, https://www.slideshare.net/vincentwambua/a-guide-to-dairy-calf
-feeding-and-management.

17. BAMN, *A Guide to Dairy Calf Feeding and Management.*

18. This is not unlike the market for human infant formula and the history of pushing formula as the "healthier" option for even those women who had no problem breast-feeding their children.

19. The website for Comfy Calf Suites, made by Canarm AgSystems, can be seen at: http://comfycalfsuites.com/ (accessed October 19, 2017).

Chapter 10

1. James Stanescu, "Species Trouble, Judith Butler, and the Precarious Lives of Animals," *Hypatia* 27, no. 3 (2012): 567–82.

2. For more on this, see Maneesha Deckha, "The Subhuman as a Cultural Agent of Violence," *Journal for Critical Animal Studies* 8, no. 3 (2010): 28–51; Claire Jean Kim, *Dangerous Crossings: Race, Species, and Nature in a Multicultural Age* (Cambridge: Cambridge University Press, 2015); Aph Ko and Syl Ko, *Aphro-ism: Essays on Pop Culture, Feminism, and Black Veganism from Two Sisters* (Brooklyn, NY: Lantern Books, 2017); Will Kymlicka and Sue Donaldson, "Animal Rights, Multiculturalism, and the Left," *Journal of Social Philosophy* 45, no. 1 (2014): 116–35.

3. Amie Breeze Harper, "Race as a 'Feeble Matter' in Veganism: Interrogating Whiteness, Geopolitical Privilege, and Consumption Philosophy of 'Cruelty-Free' Products," *Journal for Critical Animal Studies* 8, no. 3 (2010): 5–23.

4. Kymlicka and Donaldson, "Animal Rights, Multiculturalism, and the Left," 123.

5. For more information on Beacon Food Forest, an edible urban forest project in the Beacon Hill neighborhood of Seattle, WA, please visit http://beaconfoodforest.org/, and more information on Food Shift, a San Francisco Bay Area–based organization that helps route edible food destined for the landfill into local food distribution networks, is available at http://foodshift.net/.

6. Jane E. Brody, "Prescribing Vegetables, Not Pills," *Well* (blog), *New York Times* December 1, 2014, http://well.blogs.nytimes.com/2014/12/01/prescribing-vegetables-not-pills/?_r=1.

7. You can find more information about Food Not Bombs at their website: http://foodnotbombs.net/new_site/index.php.

8. For more information on Food Empowerment Project, visit http://www.foodispower.org/.

9. Sue Donaldson and Will Kymlicka, "Farmed Animal Sanctuaries: The Heart of the Movement? A Socio-Political Perspective," *Politics and Animals* 1 (2015): 50–74.

INDEX